普通高等教育“十二五”规划教材

Introduction of Professional English for Water Sciences

戴长雷　孙思淼　杜新强　刘中培　姜宁　李鹏　编

内 容 提 要

本教材由五部分构成，主要内容包括：水文循环原理，黑龙江（阿穆尔河）流域概况，地下水专业术语，重力坝、土石坝等水工建筑物的基础知识，节水灌溉方法的介绍，寒区冰雪冻土相关知识等。

全书共有33篇课文，7个图表及1个单词表，方便学生学习和掌握专业词汇，培养学生专业英语阅读与翻译能力。主要适用于水文与水资源工程、水利水电工程、农业水利工程以及相关专业的高年级本科生。

图书在版编目（CIP）数据

水科学专业英语 = Introduction of professional English for water sciences / 戴长雷等编. -- 北京 : 中国水利水电出版社，2013.12
普通高等教育“十二五”规划教材
ISBN 978-7-5170-1540-6

Ⅰ. ①水… Ⅱ. ①戴… Ⅲ. ①水文学－英语－高等学校－教材 Ⅳ. ①H31

中国版本图书馆CIP数据核字(2013)第305260号

审图号：GS（2013）2904号

书　名	普通高等教育“十二五”规划教材 **Introduction of Professional English for Water Sciences**
作　者	戴长雷　孙思淼　杜新强　刘中培　姜宁　李鹏　编
出版发行	中国水利水电出版社 （北京市海淀区玉渊潭南路1号D座　100038） 网址：www. waterpub. com. cn E-mail：sales@waterpub. com. cn 电话：（010）68367658（发行部）
经　售	北京科水图书销售中心（零售） 电话：（010）88383994、63202643、68545874 全国各地新华书店和相关出版物销售网点
排　版	北京时代澄宇科技有限公司
印　刷	北京嘉恒彩色印刷有限责任公司
规　格	184mm×260mm　16开本　9.5印张　293千字
版　次	2013年12月第1版　2013年12月第1次印刷
印　数	0001—2000册
定　价	**22.00**元

前言

作为本科教学讲义，在多次修订的基础上，《Introduction of Professional English for Water Sciences》在黑龙江大学已使用了七年。通过授课反馈、调查问卷和试卷分析等途径得知，该讲义的辅助教学效果良好。为了更有效地开展水利类专业英语教学，切实提高教学水平，2012年本教材申请获批为黑龙江大学“十二五”规划教材和中国水利水电出版社“十二五”规划教材。本教材的出版得到了“黑龙江省教育厅黑龙江省高等教育教学改革项目：寒区特色水文与水资源专业建设的研究与实践(No. 11551330)”的支持。

本教材主要涉及内容有水文循环、水利工程、灌溉排水、寒区水利以及水利专业日常用语等，主要适用于水文与水资源工程、水利水电工程、农业水利工程以及相关专业的高年级本科生。区别于其他水利类专业英语教材和讲义，《Introduction of Professional English for Water Sciences》具备以下4个特色：

(1) 强化图、表：课文编排中大量采用了图、表的表达方式，避免了冗长的课文学习，使得专业英语的学习直观、生动、形象；

(2) 专业与实用相结合：学习内容多取材于专业教材及学生日常生活、专业学习中熟知的常识、知识，可读性好；

(3) 配有页边注释及习题：每课均设有页边注释及课后习题辅助消化内容，可引导学生形成良好的专业英语学习习惯；

(4) 设有寒水特色内容：结合地域特色，在黑龙江省高等教育教学改革项目的支持下，专用Part 5设置了水情、雪情、冻土相关寒水内容。

本书稿由黑龙江大学主导组织，由黑龙江大学戴长雷副教授、姜宁讲师，硕士毕业于黑龙江大学的伯明翰大学博士研究生孙思淼，吉林大学杜新强副教授，华北水利水电大学刘中培副教授，毕业于阿拉斯加大学的长安大学老师李鹏博士共同编写完成。具体分工如下，Part 1 戴长雷、孙思淼；Part 2 刘中培、孙思淼；Part 3 姜宁、孙思淼；Part 4 李鹏、杜新强；Part 5 孙思淼、杜新强；Vocabulary 杜新强、戴长雷；全书最后由戴长雷统稿。

感谢为本书策划和出版付出辛苦劳动的朱双林编辑和魏素洁编辑，感谢为本书后期校订工作付出劳动的黑龙江大学水利电力学院研究生与本科生李欣欣、常龙艳、王思聪、商允虎、赵锡山、陈扬、银若冰、张大帅、罗章、秦国帅、牟伟楠等。

由于涉及面较广，编者的能力、时间有限，本教材尚有许多不完善之处，敬请各位老师和同学提出宝贵意见，不胜感谢。

戴长雷/daichanglei@126. com

2013年8月24日

Contents

Part 1

Figure 1. The Hydrologic Cycle

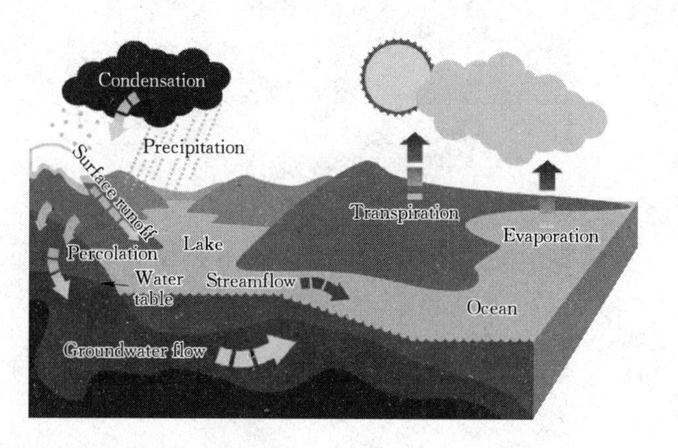

[1]hydrologic[ˌhaidrəˈlɔdʒik]
a. 水文的，水文学的
[2]precipitation[priˌsipiˈteiʃn]
n. 降水
[3]condensation [ˌkɒndenˈseiʃn]
n. 凝结，冷凝
[4]evaporation[iˌvæpəˈreiʃn]
n. 蒸发（作用）
[5]streamflow [ˌstriːmfləu] *n.* 流速及流水量
[6]runoff[ˌrʌnˌɔːf] *n.* 径流
[7]transpiration[ˌtrænspiˈreʃən]
n. 蒸发，散发；[植] 蒸腾作用；[航] 流逸

Exercises

1. Build sentences with following words.

(1) hydrologic cycle

(2) precipitation, evaporation, condensation, runoff

(3) transpiration, cease

2. Translate the following sentences into English.

(1) 在水文循环过程中，陆地上与海洋中的水通过蒸发进入大气，大气中的水凝结后形成云，经过降水作用又返回到地面和海面上。

(2) 径流分地面径流和地下径流两种，降水包括降雨、降雪、冰雹等。

3. Try to describe the process of hydrologic cycle in English.

Learning section

Inspirational[8] sayings:

Cease[9] to struggle and you cease to live. —Thomas Carlyle

生命不止，奋斗不息。——托马斯·卡莱尔

[8]inspirational[ˌinspəˈreʃənl]
a. 鼓舞人心的；带有灵感的，给予灵感的
[9]cease[siːs] *vi.* 停止；终了 *vt.* 停止；结束 *n.* 停止

Living without an aim is like sailing without a compass. —John Ruskin
生活没有目标，犹如航海没有罗盘。——罗斯金
A bold attempt is half success.
勇敢的尝试是成功的一半。
God helps those who help themselves.
天助自助者。

Figure 2. Urban Rainfall Capture System

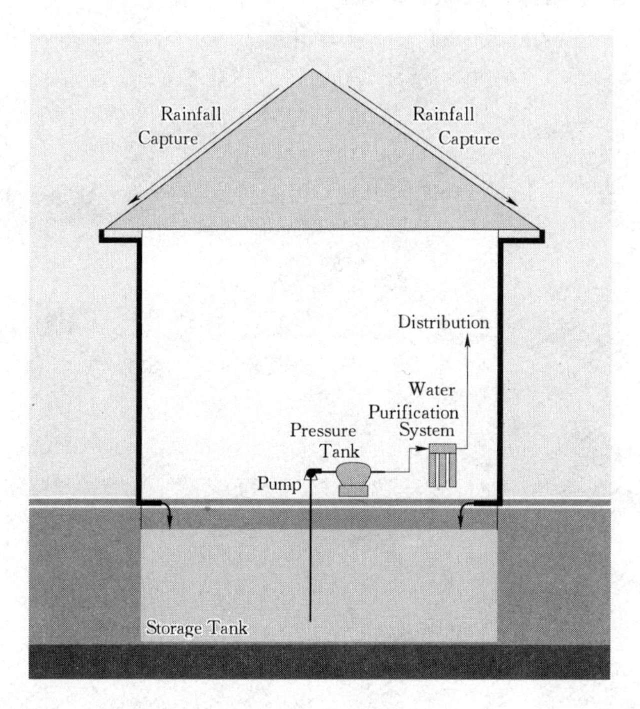

[1]urban ['ɜːbən] *a.* 城市的

[2]distribution[ˌdistri'bjuːʃn] *n.* 分发，分配；散布，分布

[3]purification[ˌpjʊərifi'keiʃn] *n.* 净化

[4]pressure tank 压力槽，高压锅，耐压试验水筒，坞室

[5]pump[pʌmp] *n.* 泵，打气筒 *vt.* & *vi.* 用抽水机汲水，给……打气

[6]storage tank 储水罐

Exercises

1. Build sentences with following words.

(1) urban rainfall，system

(2) purification，polluted water

(3) pump，storage tank

2. Translate the following sentences into English.

(1) 雨水收集器装置、储水罐、水泵和净化设备都是城市雨水收集系统的重要组成部分。

(2) 城市雨水收集系统可将雨水收集到储水罐中，通过压力槽和净化装置后使用。

Learning section

You can lead a horse to water，but you can't make him drink.

强扭的瓜不甜。

Water over the dam.

木已成舟，既成事实。

He knows the water the best who has **waded**[7] through it.
要知河深浅，须问过来人。
Smooth waters run deep.
静水流深；沉默者深谋。

[7]wade[weid]*vt.* & *vi.*（从水、泥等）蹚，走过；跋涉 *n.* 跋涉

Unit 1. The Hydrologic Cycle

Passage

In nature, water is constantly changing from one state to another. The heat of the sun **evaporates**[1] water from land and water surfaces. This water **vapor**[2] (a gas), being lighter than air, rises until it reaches the cold upper air where it **condenses**[3] into clouds. Clouds **drift**[4] around according to the direction of the wind until they strike a colder atmosphere. At this point the water further condenses and falls to the earth as rain, **sleet**[5], or snow, thus completing the **hydrologic**[6] cycle.

The complete hydrologic cycle, however, is much more complex. The atmosphere gains water vapor by evaporation not only from the oceans but also from lakes, rivers and other water bodies, and from **moist**[7] ground surfaces. Water vapor is also gained by **sublimation**[8] from snowfields and by **transpiration**[9] from **vegetation**[10] and trees.

Water **precipitation**[11] may follow various **routes**[12]. Much of the precipitation from the atmosphere falls directly on the oceans. Of the water that does fall over land areas, some is caught by vegetation or evaporates before reaching the ground, some is locked up in snowfields or ice-fields for periods ranging from a season to many thousands of years, and some is **retarded**[13] by **storage**[14] in **reservoirs**[15], in the ground, in chemical compounds, and in vegetation and animal life.

The water that falls on land areas may return immediately to the sea as **runoff**[16] in **streams**[17] and rivers or when snow melts in warmer seasons. When the water does not run off immediately it **percolates**[18] into the soil. Some of this **groundwater**[19] is taken up by the **roots**[20] of vegetation and some of it flows through the **subsoil**[21] into rivers, lakes, and oceans.

Because water is absolutely necessary for **sustaining**[22] life and is of great importance in industry, men have tried in many ways to control the hydrologic cycle to their own advantage. An obvious example is the storage of water behind dams in reser-

[1] evaporate [i'væpəreit] *vt.* & *vi.* *vt.*（使某物）蒸发掉 *vi.* 消失，不复存在

[2] vapor ['veipə] *n.* 水汽，水蒸气 *v.* 自夸；（使）蒸发

[3] condense [kən'dens] *v.* 浓缩

[4] drift [drift] *vi.* 漂，漂流；漂泊，流浪 *n.* 漂移，漂流

[5] sleet [sli:t] *n.* 雨夹雪或雹 *vi.* 下雨夹雪；下冻雨

[6] hydrologic ['haidrəulɔdʒik] *a.* 水文的；水文学的 hydrologic cycle 水文循环

[7] moist [mɔist] *a.* 潮湿的，微湿的，湿润的

[8] sublimation [ˌsʌbli'meiʃn] *n.* 升华，升华作用

[9] transpiration [ˌtrænspi'reiʃn] *n.* 蒸发（物），散发，蒸腾作用，流逸

[10] vegetation [ˌvedʒə'teiʃn] *n.* 植物（总称）；草木

[11] precipitation [priˌsipi'teiʃn] *n.* 降水

[12] route [ru:t] *n.* 路程，道路；途径；渠道槽

[13] retard ['ri:tɑ:d] *vt.* 停滞 *vi.* 减慢；受到阻滞 *n.* 减速；阻滞；延迟

[14] storage ['stɔ:ridʒ] *n.* 贮存，贮藏

[15] reservoir ['rezəvwɑ:(r)] *n.* 水库；储藏，汇集；大量的储备；储液

[16] runoff [rʌnɔf] *n.* 径流

[17] streams [st'ri:mz] *n.* 流（stream 的名词复数）；一连串；水流方向；小河

[18]percolate[ˈpɜːkəleit]vt. 渗滤 vi.（思想等）渗透，渗入
[19]groundwater [ˈgraʊndwɔːtə(r)] n. 地下水
[20]root [ruːt] n. 根，根部
[21]subsoil [ˈsʌbsɔil]n. 下土层，底土
[22]sustain [səˈstein] vt. 维持，使……生存下去
[23] excess [ikˈses] n. 过量，过剩
[24] deficit [ˈdefisit] n. 欠缺，不足
[25] inject [inˈdʒekt] vt. 注射，喷射
[26] particle [ˈpɑːtikl] n. 微粒，颗粒，[物] 粒子
[27]iodide [ˈaiədaid] n. 碘化物
[28] modification [ˌmɒdifiˈkeiʃn] n. 改善，改变
[29]meteorologist [ˌmiːtiəˈrɒlədʒist] n. 气象学家
[30]contour [ˈkɒntʊə(r)] n. 外形，轮廓，等高线 a. 沿等高线修筑的
[31]plowing [plauiŋ] n. 耕地
[32]sloping [ˈsləʊpiŋ] a. 倾斜的，有坡度的
[33]dike[daik] n. 堤，坝
[34]purification [ˌpjʊərifiˈkeiʃn] n. 净化
[35]predict[priˈdikt] vt. 预测
[36]intensity [inˈtensəti] n. 强度
[37]watershed [ˈwɔːtəʃed] n. 流域
[38] hydrologist [haiˈdrɒlədʒist] n. 水文学家

voirs, in climates where there are **excesses**[23] and **deficits**[24] of precipitation (with respect to water needs) at different times in the year. Another method is the attempt to increase or decrease natural precipitation by **injecting**[25] **particles**[26] of dry ice or silver **iodide**[27] into clouds. This kind of weather **modification**[28] has had limited success thus far, but many **meteorologists**[29] believe that a significant control of precipitation can be achieved in the future.

Other attempts to influence the hydrologic cycle include the **contour**[30] **plowing**[31] of **sloping**[32] farmlands to slow down runoff and permit more water to percolate into the ground, the construction of **dikes**[33] to prevent floods and so on. The reuse of water before it returns to the sea is another common practice. Various water supply systems that obtain their water from rivers may recycle it several times (with **purification**[34]) before it finally reaches the rivers mouth.

Men also attempt to **predict**[35] the effects of events in the course of the hydrologic cycle. Thus, the meteorologist forecasts the amount and **intensity**[36] of precipitation in a **watershed**[37], and the **hydrologist**[38] forecasts the volume of runoff.

(710 words)

Exercises

1. Build sentences with following words.

(1) hydrologic cycle, transpiration, precipitation, percolate
(2) vapor, condense, sleet
(3) watershed, runoff, drought
(4) storage, reservoir
(5) sublimation, evaporate

2. Translate the following sentences into English.

(1) 水文循环的三要素是降水、蒸发与径流。

(2) 水文循环分为小循环与大循环。海洋与大陆之间的水分交换称为大循环，海洋或大陆内部的水分交换称为小循环。

(3) 水文循环现象发生的原因之一是水在常温下就能实现液态、气态和固态的三态转换而不发生化学变化。

3. Try to describe the process of hydrologic cycle according to the figure below (Fig. 1. 1).

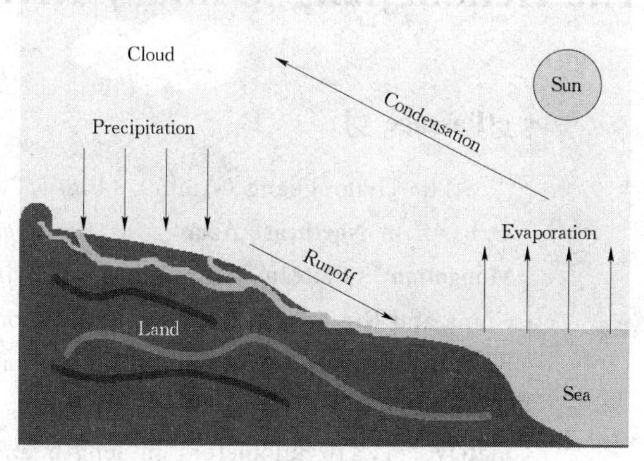

Fig. 1. 1 The process of hydrologic cycle

Learning section

Tab. 1. 1 **Common Abbreviations 1**

星期		
Mon.	Monday	星期一
Tues.	Tuesday	星期二
Wed.	Wednesday	星期三
Thur.	Thursday	星期四
Fri.	Friday	星期五
Sat.	Saturday	星期六
Sun.	Sunday	星期日

Unit 2. The Heilongjiang (Amur) River Basin

Passage

The Heilongjiang (Amur) (Fig. 1. 2) is the largest river basin in **Northeast Asia**[1]. It flows eastwards from the **Mongolian**[2] **Plateau**[3] through 13 provinces of Mongolia, China and Russia and covers a tiny bit on **North Korea**[4] at Songhua River **headwaters**[5]. The Heilong (Amur) River is one of the world's largest free - flowing rivers at **approximately**[6] 4, 416 kilometers in length and its watershed is 1. 83 million square kilometers. The river forms the border between China and Russia for over 3,000km, making it one of the world's longest border rivers.

Heilongjiang (Amur) River Basin is largely mountainous. Mountain ranges, ridges, **foothills**[7] and plateaus cover two - thirds of the region. **Henty**, **Great Hinggan**, **Stanovoy**[8], Tukuringra, Bureinsky, Sikhote - Alin, Changbaishan, Small Hinggan are principal mountain ranges of the area. Most mountains are low and covered with forest, ranging from 300 to 1,000 meters in **elevation**[9]. Only isolated mountain ranges and peaks, covering just over seven percent of the **territory**[10], reach elevations exceeding 2,000 meters. The area covered by plains is also large. **Extensive**[11], hilly **Daurian Steppe Plateau**[12] occupies the south - west part of the basin in Mongolia, Russia and Inner Mongolia. The main **plains**[13] and **lowlands**[14] are located between the **Zeya**[15] and **Bureya**[16] Rivers, near **Khanka**[17] Lake, in the valley of the Lower Amur, at the **confluence**[18] of the Heilongjiang (Amur) with Songhua and **Ussuri**[19] Rivers (Sanjiang Plain), and in the middle reaches of the Songhua at the confluence of the Nen and Second Songhua Rivers, which **drain**[20] the Song - Nen Plain.

The basin is rich in biological diversity and supports thousands of species and many ecosystem types. This vast area is famous for rare **cranes**[21] and **storks**[22], tigers and

[1]Northeast Asia 东北亚
[2]Mongolian[mɔŋ'guliən] *n.* 属于蒙古人种的人，蒙古语，蒙古症患者
[3]plateau ['plætəu] *n.* 高原，平稳，稳定状态
[4]North Korea 朝鲜
[5]headwater ['hedwɔːtə(r)] *n.* 上游源头
[6] approximately [ə'prɒksimətli] *ad.* 近似地，大约
[7]foothill ['futhil] *n.* 山麓小丘
[8]Henty 亨提山，Great Hinggan 大兴安岭，Stanovoy 外兴安岭
[9]elevation [ˌeli'veiʃn] *n.* 提拔，海拔，提高
[10]territory ['teritəri] *n.* 领土，版图，领域，范围
[11]extensive [iks'tensiv] *a.* 广泛的，广阔的
[12]Daurian Steppe Plateau 达乌尔草原高原
[13]plain [plein] *n.* 平原
[14]lowland ['ləulænd]*n.* 低地
[15]Zeya *n.* 泽亚河
[16]Bureya *n.* 布列亚河
[17]Khanka *n.* 兴凯湖
[18]confluence ['kɒnfluəns] *n.* 合流，合流点，集合
[19] Ussuri [u'suːri] *n.* 乌苏里江
[20]drain [drein] *n.* 下水道，排水沟，消耗 *v.* 耗尽，排出沟外
[21]crane [krein] *n.* 鹤，起重机 *v.* 引颈，伸长（脖子）*vt.* 伸长（脖子等）
[22]stork [stɔːk] *n.* [鸟]鹳

leopards[23], and **endemic**[24] fishes. The biological richness is explained by a great diversity of landscapes such as floodplain wetlands, **steppe**[25], **alpine**[26] **tundra**[27], mixed broadleaf - coniferous forest, and **boreal**[28] **taiga**[29]. Approximately 10% of the basin territory is covered by protected areas, and some are **nominated**[30] as **Ramsar wetlands**[31], **UNESCO**[32] **biosphere reserves**[33], World **heritage**[34] sites, etc. Despite **uniqueness**[35] and **grandeur**[36] of many natural areas they are little known by outside world and have not become international tourism attractions. So far **national nature conservation programs**[37] and **international environmental agreements**[38] and efforts are insufficient and cannot prevent rapid **deterioration**[39] of environment and biodiversity loss in Amur basin.

Degree of human impact on the environment in Amur River basin is uneven, but already quite substantial in all countries. Plains support extensive agriculture, especially in China, while other dominant land - use types are **extraction**[40] of **mineral**[41] resources, and forestry. Development of international trade spurs construction of transportation infrastructure, oil & gas pipelines. Several major tributaries are already blocked by hydropower dams, while lower reaches are severely affected by water pollution from industry and agriculture. Excessive harvest of biological resources: **timber**[42], fish, **terrestrial**[43] wildlife is also **triggered**[44] by international market demand. Poor land - management practices and accelerating climate change lead to widespread wildfires and land **degradation**[45].

Heilongjiang (Amur) River basin exemplifies transboundary regions in need of shared environmental responsibility. Land - use patterns and cultural traditions and pace of economic development are drastically different in Russia, Mongolia and China, but sustainable development requires cooperation in the field of environmental protection and nature resource management.

(901 words)

[23]leopard [ˈlepəd] *n.* 豹

[24]endemic [enˈdemik] *n.* 地方病，风土病 *a.* 风土的，地方的 endemic fish 特有鱼种

[25]steppe [step] *n.* 特指西伯利亚一带没有树木的大草原

[26]alpine [ˈælpain] *a.* 高山的

[27]tundra [ˈtʌndrə] *n.* 苔原，冻土地带

[28]boreal [bɔriəl] *a.* 北方的

[29]taiga [ˈteigə] *n.* 针叶树林地带

[30]nominated [ˈnɔmineitid] *a.* 被提名的，被任命的

[31]Ramsar wetlands 拉姆萨尔湿地（《拉姆萨尔公约》意为国际重要湿地）

[32]UNESCO—United Nations Educational, Scientific and Cultural Organization [juːˈneskəʊ] 联合国教育科学文化组织

[33]biosphere reserves 生物保护圈

[34]heritage [ˈheritidʒ] *n.* 遗产，继承物

[35]uniqueness[juːˈniːknis]*n.* 独特性

[36]grandeur[ˈgrændjə]*n.* 富丽堂皇

[37]national nature conservation programs 国家自然保护项目

[38]international environmental agreements 国际环境协定

[39]deterioration[diˌtiəriəˈreiʃn] *n.* 恶化，退化，变坏

[40]extraction [ikˈstrækʃn] *vt.* 提取

[41]mineral [ˈminərəl] *a.* 矿物的 *n.* 矿物，矿石

[42]timber [ˈtimbə] *n.* 木材，木料

[43]terrestrial [tiˈrestriəl] *n.* 地球上的人

[44]triggered [ˈtrigəd] *a.* 触发的，起动的

[45]degradation [ˌdegrəˈdeiʃn] *n.* 降格，堕落，退化

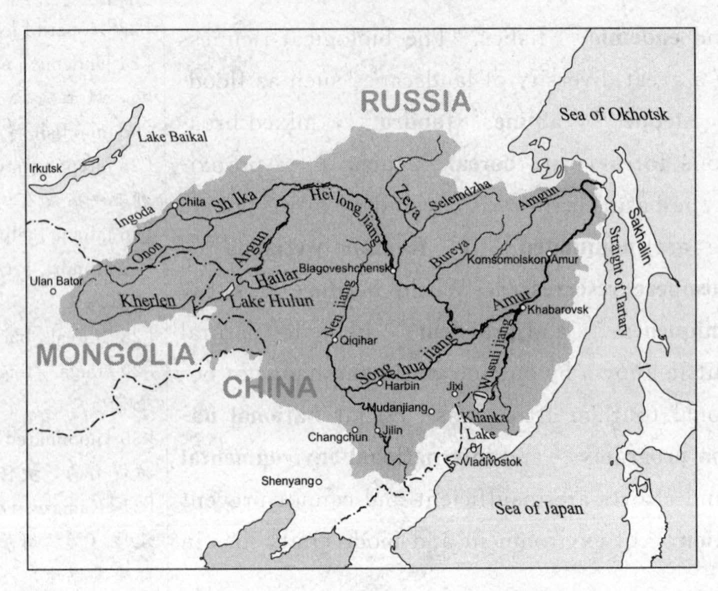

Fig. 1. 2 Map of theHeilongjiang (Amur) River

Exercises

1. Build sentences with the following words.

(1) basin, main stream, tributary

(2) hydropower, dam, lock

(3) Pacific, Arctic, Atlantic, Indian

(4) Northeast Asia, North Korea, Khanka

2. Translate the following sentences into English.

(1) 黑龙江（阿穆尔河）是世界第 8 长河，也是仅次于长江、黄河的中国第 3 长河。

(2) 黑龙江（阿穆尔河）有 7 条主要的支流，它们是中国的额尔古纳河、松花江、乌苏里江和俄罗斯的石勒喀河、泽亚河、布里亚河和阿姆干河。

(3) 一条河的主要特征参数包括：河长、位置、流向、流量和流域面积。

(4) 黑龙江（阿穆尔河）全长 4416km，注入北太平洋鄂霍茨克的鞑靼海峡。

(5) 黑龙江（阿穆尔河）在 5 月中旬到 11 月中旬约有5～6个月的冰期，河上没有水电大坝、船闸和大的工厂。

(6) 黑龙江（阿穆尔河）沿岸有 9 个重要的城市，其中中国 5 个，俄罗斯 4 个。

(7) 黑龙江（阿穆尔河）流域是东北虎、亚洲白鹳和丹顶鹤的栖息地。

Learning section

Tab. 1. 2 **Common Abbreviations 2**

月份		
Jan.	January	1月
Feb.	February	2月
Mar.	March	3月

续表

月份		
Apr.	April	4月
May	May	5月
Jun.	June	6月
Jul.	July	7月
Aug.	August	8月
Sep.	September	9月
Oct.	October	10月
Nov.	November	11月
Dec.	December	12月

Unit 3. Global Water Reservoirs and Fluxes[1]

[1] flux [flʌks] *n.* 通量，焊剂 *vi.* 流动 *vt.* 出水
[2] virtually [ˈvɜːtʃuəli] *ad.* 事实上
[3] accessible [əkˈsesəbl] *a.* 允许的，可进入的，可达的
[4] glacier [ˈglæsiə(r)] *n.* 冰川，冰河
[5] distribution [ˌdistriˈbjuːʃn] *n.* 分配，周延性，分布
[6] glacial ice 冰川冰
[7] accessibility [əkˌsesəˈbiləti] *n.* 可达性，可及性，可及度
[8] groundwater volume 地下水容积
[9] crust [krʌst] *n.* 外壳，外皮，地壳
[10] saline [ˈseilain] *n.* 盐泉，盐水，盐湖 *a.* 含盐的，盐的，苦涩的
[11] marsh [mɑːʃ] *n.* 沼泽，湿地，草沼
[12] biosphere [ˈbaiəusfiə(r)] *n.* 生物圈，生命层，生物界
[13] solar radiation 太阳辐射

Passage

Water exists in **virtually**[2] every **accessible**[3] environment on or near the earth's surface, it's in blood, trees, air, **glaciers**[4], streams, lakes, oceans, rocks, and soil. The total amount of water on the planet is about 1.4×10^9 km^3, and its **distribution**[5] among the main reservoirs is listed in Tab. 1.3 (Maidment, 1993). Of the fresh water reservoirs, **glacial ice**[6] and groundwater are by far the largest. Groundwater and surface water are the two reservoirs most used by humans because of their **accessibility**[7]. Fresh groundwater is about 100 times more plentiful than fresh surface water, but we use more surface water because it is so easy to find and use. Much of the total **groundwater volume**[8] is deep in the **crust**[9] and too saline for most uses.

Tab. 1.3 Distribution of Water in Earth's Reservoirs

Resevoir	Percent of All Water	Percent of Fresh Water
Oceans	96.5	
Ice and snow	1.8	69.9
Groundwater		
Fresh	0.76	30.1
Saline	0.93	
Surface Water		
Fresh lakes	0.007	0.26
Saline[10] lakes	0.006	
Marshes[11]	0.0008	0.03
Rivers	0.0002	0.006
Soil moisture	0.0012	0.05
Atmosphere	0.001	0.04
Biosphere[12]	0.0001	0.003

Source: Maidment (1993).

Fueled by energy from **solar radiation**[13], water changes phase and cycles continuously among these reservoirs in the hydrologic cycle (Fig. 1. 3). **Solar energy**[14] **drivers**[15] evaporation, transpiration, **atmospheric circulation**[16], and precipitation. Gravity pulls precipitation down to earth and pulls surface water and groundwater down to lower elevations and **ultimately**[17] back to the ocean reservoir. Evaporation and transpiration are difficult to measure **separately**[18], so their combined effects are usually **lumped**[19] together and called evapotranspiration.

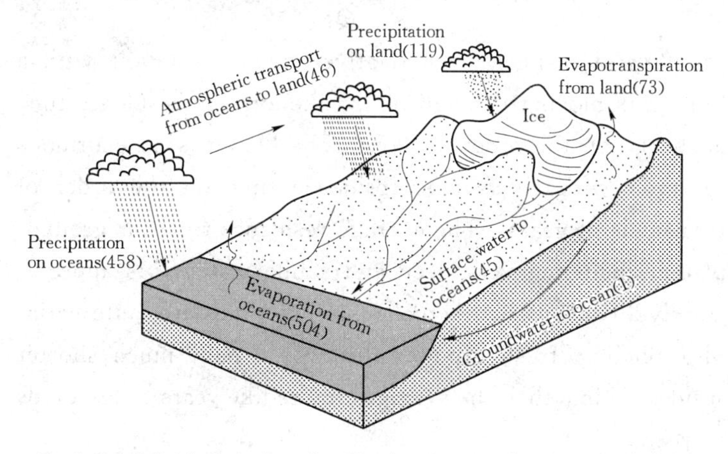

Fig. 1. 3 Global hydrologic cycle. Numbers in parentheses are total global fluxes in thousands of km^3/yr. Data from Maidment (1993)

Over land areas, average precipitation exceeds average evapotranspiration. The opposite is true over the oceans. On average, more atmospheric water moves from the ocean areas to land areas than **vice versa**[20], creating a **net flux**[21] of atmospheric water from ocean areas. The flux of surface water and groundwater from the land back to the oceans maintains a balance so that the volumes in each reservoir remain **roughly**[22] constant over time. The hydrologic cycle represents only global averages; the actual fluxes in smaller regions and smaller time **frames**[23] **deviate**[24] **significantly**[25] from the average. Deserts, for example, are **continental**[26] areas where evaporation exceeds precipitation. On the other hand, at a cold, rainy **coastline**[27] like the northwest Pacific, precipitation exceeds evaporation.

In a given region, the fluxes are distributed **irregularly**[28] in time due to specific storm events or seasonal variations such as **monsoons**[29]. With these transient fluxes, the

[14]solar energy 太阳能
[15]drivers[d'raivəz]*n.* 驱动
[16] atmospheric circulation 大气循环
[17]ultimately['ʌltimətli] *ad.* 最后
[18]separately['seprətli] *ad.* 独立地
[19]lumped[lʌmpt] *a.* 集总的 *vt.* 集总

[20]vice versa 反之亦然
[21]net flux 流网
[22]roughly['rʌfli]*ad.* 概略地
[23]frame[freim] *n.* 帧，框架，架
[24]deviate['di:vieit] *vi.* 偏离，背离，偏差
[25] significantly [sig 'nifikəntli] *ad.* 意义重大地
[26]continental[ˌkɒnti'nentl] *a.* 大陆的 *n.* 大陆人
[27]coastline['kəʊstlain] *n.* 海岸线
[28]irregularly[i'regjələli] *a.* 不整齐的，不规则的 *n.* 不合规格之物
[29]monsoon[ˌmɒn'su:n] *n.* 季风，季节风

reservoir volumes **fluctuate**[30]; groundwater and surface water levels rise and fall, glaciers grow and **shrink**[31], and sea level rises and falls slightly.

[30] fluctuate['flʌktʃueit] *vi.* 波动，起伏 *vt.* 拨动

[31] shrink[ʃriŋk] *vt.* & *vi.* 收缩，皱缩

[32] residence time 停留时间

The **residence time**[32] is the average amount of time that a water molecule resides in a particular reservoir before transferring to another reservoir. The residence time T_r is calculated as the volume of a reservoir V[length^3 or L^3] divided by the total flux in or out of the reservoir Q[length^3 per time or L^3]

$$T_r = \frac{V}{Q} \tag{1.1}$$

The atmosphere is a relatively small reservoir with a large flux moving through it, so the average residence time is short, on the order of days. The Ocean is an enormous reservoir with an average residence time on the order of thousands of years. The average residence time for groundwater, including very deep and saline waters, is approximately 20,000years. Actual residence times are quite variable. Shallow fresh groundwater would have much shorter residence time than the average, more like years to hundreds of years.

(619 words)

Exercises

1. Translate the following sentences into Chinese.

(1) Fresh groundwater is about 100 times more plentiful than fresh surface water, but we use more surface water because it is so easy to find and use.

(2) The flux of surface water and groundwater from the land back to the oceans maintains a balance so that the volumes in each reservoir remain roughly constant over time.

(3) More atmospheric water moves from the ocean areas to land areas than vice versa, creating a net flux of atmospheric water from ocean areas.

(4) With these transient fluxes, the reservoir volumes fluctuate; groundwater and surface water levels rise and fall, glaciers grow and shrink, and sea level rises and falls slightly.

(5) The residence time is the average amount of time that a water molecule resides in a particular reservoir before

transferring to another reservoir.

2. Translate the following sentences into English.

(1) 地球上水的总量约为 1.4×10^9 km^3，它在主要水库中的分布情况列于表中。

(2) 太阳能驱动了蒸发、蒸腾、大气循环和降水。

(3) 蒸发和蒸腾很难单独测量，所以通常将其联合作用视为一体，称为蒸散发。

(4) 水循环仅仅代表全球的平均水平，在一些较小地区和时间尺度内，实际的流量与平均水平有较大偏差。

(5) 沙漠是蒸发量大于降水量的大陆地区。

Learning section

Tab. 1. 4　　Common Abbreviations 3

日常缩写		
VIP	very important person	贵宾
IMP	import	进口
EXP	export	出口
MAX	maximum	最大
MIN	minimum	最小
DOC	document	文件
Tel.	Telephone number	电话号码
appx.	appendix	附录
av.	average	平均
B. C	before Christ	公元前
A. C.	alternating current	交流（电）
D. C.	direct current	直流（电）
vol.	volume	体积，卷
Wt	weight	重量

Unit 4. Terminology for Subsurface Waters

Passage

[1]branch[brɑːntʃ] *n.* 树枝；分支；部门，分科；支流
[2]terminology[ˌtəːmiˈnɒlədʒi] *n.* 专门名词；术语，术语学
[3]categorization[ˌkætəgəraiˈzeiʃn] *n.* 编目方法，分门别类
[4]adopt[əˈdɒpt] *vt.* 采用，采取，采纳

No **branch**[1] of science is without its **terminology**[2]. Before going further, we must define the terms used to discuss subsurface waters. The traditional **categorization**[3] of subsurface water, and the one **adopted**[4] here, is shown in Fig. 1. 4.

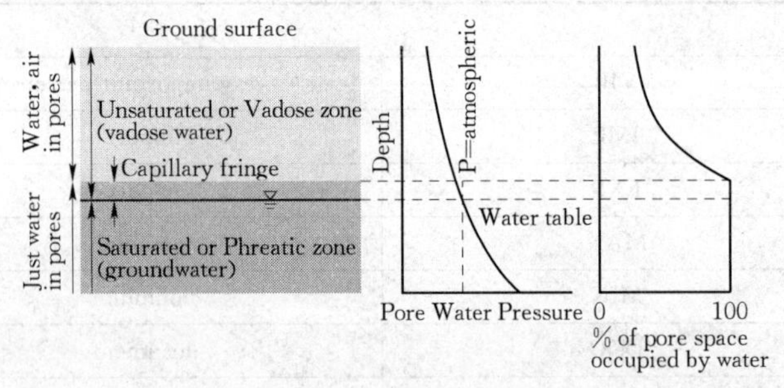

Fig. 1. 4 Vertical cross－section showing the definitions of terms used to describe subsurface water

[5]unsaturated[ˈʌnˈsætʃəreitid] *a.* 没有饱和的，不饱和的
[6]vadose zone 渗流区
[7]saturated[ˈsætʃəreitid] *a.* 饱和的；浸透的；（颜色）未被白色弄淡的
[8]phreatic zone 地下水层；地下水区；潜水带；饱和层
[9]water table 地下水位
[10]technically[ˈteknikli] *ad.* 技术上；学术上；专业上；严格说来
[11]pore water pressure 孔隙水压力
[12]synonymous[siˈnɒniməs] *a.* 同义的；同义词的
[13]stabilize[ˈsteibilaiz] *vt.* 稳定 *vi.* 安定
[14]capillary force 毛管力
[15]mineral surfaces 矿物表面

Subsurface waters are divided into two main categories: the near－surface **unsaturated**[5] or **vadose zone**[6] and the deeper **saturated**[7] or **phreatic zone**[8]. The boundary between these two zones is the **water table**[9], which is **technically**[10] defined as the surface on which the **pore water pressure**[11] equals atmospheric pressure. In cross－section drawings like Fig. 1. 4, the water table and other water surface are typically marked with the symbol $\underline{\nabla}$. The terms phreatic surface and free surface are **synonymous**[12] with water table. Measuring the water table is easy. If a shallow well is installed so it is open just below the water table, the water level in the well will **stabilize**[13] at the level of the water table.

The unsaturated zone or vadose zone is defined as the zone above the water table where the pore water pressure is less than atmospheric. In most of the unsaturated zone, the pore spaces contain some air and some water. **Capillary forces**[14] attract water to the **mineral surfaces**[15], causing water pressures to be less than atmospheric. The term **soil**

water[16] applies to water in the unsaturated zone, usually **in reference to**[17] water in the shallow part where plant roots are active.

Below the water table is the saturated zone or phreatic zone, where water pressures are greater than atmospheric and the pores are saturated with water. Groundwater is the term for water in the saturated zone. **Aquifer**[18] is a familiar term, meaning a **permeable**[19] region or layer in the saturated zone. This book deals with both vadose water and groundwater, but most of the emphasis is on groundwater, since it is the main reservoir of subsurface water.

[16]soil water 土壤水
[17]in reference to 关于
[18]aquifer['ækwifə(r)]n. 地下蓄水层，砂石含水层
[19]permeable['pɜːmiːəbəl] a. 可渗透的，具渗透性的

The **capillary fringe**[20] is a zone that is saturated with water, but above the water table. It has traditionally been **assigned to**[21] the unsaturated zone, even though it is physically continuous with and similar to the saturated zone. The thickness of the capillary fringe varies depending on the pore sizes in the **medium**[22]. Media with small pore sizes have thicker capillary fringes than media with larger pore sizes. In **silt**[23] or clay, the capillary fringe can be more than a meter thick, while the capillary fringe in **coarse gravel**[24] would be less than a **millimeter**[25] thick. In **finer - grained**[26] materials, there is more surface area and the greater overall surface attraction forces **result in**[27] a thicker capillary fringe.

(438 words)

[20]capillary fringe 毛细管条纹
[21]assign to 把……分配给
[22]medium['midiəm] n. 媒质，介质
[23]silt[silt]n. 淤泥，泥沙沉积，粉砂
[24]coarse gravel 粗砾，粗砾石，粗卵石
[25]millimeter['miliˌmiːtə] n. 毫米
[26]finer - grained 细粒的
[27]result in 引起

Exercises

1. Build sentences with the following words.

(1) water table, technically defined, pore water pressure, atmospheric pressure

(2) Groundwater, saturated zone, aquifer, permeable region

2. Translate the following sentences into Chinese.

(1) Subsurface waters are divided into two main categories: the near - surface unsaturated or vadose zone and the deeper saturated or phreatic zone.

(2) The boundary between these two zones is the water table, which is technically defined as the surface on which the pore water pressure equals atmospheric pressure.

(3) If a shallow well is installed so it is open just below the water table, the water level in the well will stabilize at the level of the water table.

(4) Below the water table is the saturated zone or phreatic zone, where water pressures are greater than atmospheric and the pores are saturated with water.

(5) In silt or clay, the capillary fringe can be more than a meter thick, while the capillary fringe in coarse gravel would be less than a millimeter thick.

3. Translate the following sentences into English.

(1) 任何学科都有其自身的专业术语。

(2) 潜水面、自由水面这两个术语与地下水位是同义的。

(3) 不饱和带或渗流区被定义为地下水位以上的区域，这一区域的孔隙水压力小于大气压力。

(4) 地下水这一术语是指饱和带中的水。地下蓄水层是一个与之相似的术语，意为饱和带中具有渗透性的区域或地层。

(5) 毛细管条纹的厚度随介质中空隙大小而变化。

Learning section

Tab. 1. 5　　Common Abbreviations 4

国家		
FRA	France	法国
GER	Germany	德国
GBR	Great Britain	英国
IND	India	印度
ITA	Italy	意大利
JPN	Japan	日本
NED	Netherlands	荷兰
PRK	People's Republic of Korea	朝鲜
NOR	Norway	挪威
RUS	Russia	俄罗斯
CHN	China	中国
USA	United States of America	美国

Unit 5. Fluxes Affecting Groundwater

Passage

Water fluxes in and near the subsurface illustrated **schematically**[1] in Fig. 1. 5. Precipitation events bring water to the land surface, and from there water can do one of three things. It can **infiltrate**[2] into the ground to become infiltration, it can flow across the ground surface as **overland flow**[3], or it can evaporate from the surface after the precipitation stops. Water in the unsaturated zone that moves downward and flows into the saturated zone is called **recharge**[4]. Water in the unsaturated zone that flows laterally to a surface water body is known as **interflow**[5]. Some groundwater **discharges**[6] from the saturated zone back to surface water bodies. Groundwater can also exit the saturated zone as transpiration or **well discharge**[7].

Infiltration and Recharge

Whether water on the ground surface infiltrates or becomes overland flow depends on several factors. Infiltration is favored where there is **porous and permeable soil**[8] or rock, **flat topography**[9], and a history of dry conditions. Urban and suburban development creates many **impermeable**[10] surface—roofs, pavement, and **concrete**[11]—and increases overland flow at the expense of infiltration. The increased fraction of precipitation going to overland flow often leads to more frequent flooding events in urbanized areas (Leopold, 1994).

During a large precipitation event, the first water to arrive is easily infiltrated, but later in the event the pores of **surficial**[12] soils become more saturated and the rate of infiltration slows. When the rate of precipitation exceeds the rate that water can infiltrate, water will begin to **puddle**[13] on the surface. Puddles can only store so much water and then they **spill over**[14], **contributing to**[15] overland flows as shown in Fig. 1. 5. Infiltration can continue beyond the end

[1]schematic[skiː'mætik]*a.* 纲要的，示意的，概要的

[2]infiltrate['infiltreit] *vt.* & *vi.* （使）渗透，（使）渗入

[3] overland flow 表面径流，漫流

[4]recharge[riː'tʃɑːdʒ]*vt.* 再充电，再控告 *n.* 再充电，补给

[5]interflow[ˌintə(ː)'ʃləu] *n.* 混流，层间流，土内水流 *vi.* 混流，合流

[6]discharge[ˌdis 'tʃɑːdʒ] *n.* 出院，流量，放射 *vi.* 放电，流出 *vt.* 清偿，排放，放出

[7]well discharge 井流量

[8]porous and permeable soil 多孔的（具导管的，有孔的）渗透性土壤

[9]flat topography 地势平坦

[10]impermeable[im'pɜːmiːəbəl] *a.* 不能渗透的，不透水，不渗透的

[11]concrete[kənkviːt] *a.* 具体的，基本的 *n.* 凝结物，愈合，凝香体 *vt.* 凝结，使固结

[12] surficial[sɜː'fiʃəl] *a.* 地表的，地面的

[13]puddle['pʌdl] *n.* 水坑，熔池 *vi.* 和泥浆

[14]spill over 从……溢出

[15]contribute to 导致

of precipitation as puddled water drains into the subsurface.

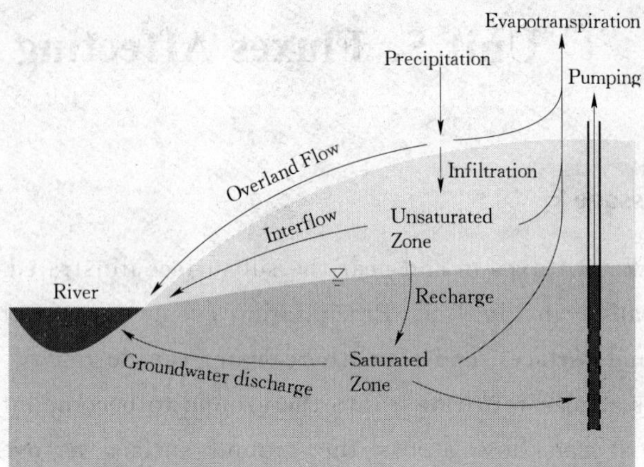

Fig. 1. 5 Reservoirs (large type) and water fluxes (small type) affecting groundwater

Rates of precipitation, infiltration, overland flow, recharge, and other fluxes illustrated in Fig. 1. 6 are often reported with units like cm/year or inches/hour[L/T]. These rates actually represent volume of water per time per area of land surface area[$L^3/T \cdot L^2$], which **reduces to**[16] length per time[L/T].

Water that infiltrates the subsurface is pulled downward by gravity, but may be **deflected**[17] **horizontally**[18] by low-permeability layers in the unsaturated zone. Horizontal flux in the unsaturated zone is called interflow. Water that moves from the unsaturated zone down into the saturated zone is called recharge. Where the unsaturated zone is thin compared to the distance to the nearest surface water, most infiltrated water becomes recharge and in such cases interflow is often neglected. Interflow becomes more significant **adjacent to**[19] stream banks and other surface waters.

Recharge is highest in areas with wet climates and permeable soil or rock types. In permeable materials, the rate of recharge can be as much as half the precipitation rate, with little overland flow, an example of such a **setting**[20] is sandy **Cape Cod**[21], **Massachusetts**[22], where the average precipitation rate is about 46 inches/year and the average recharge rate is about 21 inches/year (LeBlanc, 1984). On the other hand, in low-permeability materials only a small fraction of the precipitation becomes recharge. With massive

[16]reduce to 把……降低到
[17]deflect[di'flekt] *vi.* 偏转，偏斜 *vt.* 使偏斜
[18]horizontally[hɔri'zɔntəli] *ad.* 水平置中
[19]adjacent to 与……毗连的
[20]setting['setiŋ] *n.* 背景，装置，栽植
[21]Cape Cod 科德角
[22]Massachusetts *n.* 马萨诸塞州

clay soils[23], the recharge rate can be less than 1% of the precipitation rate.

[23]clay soils 黏性土壤
[24]nonpumping well 非抽水井

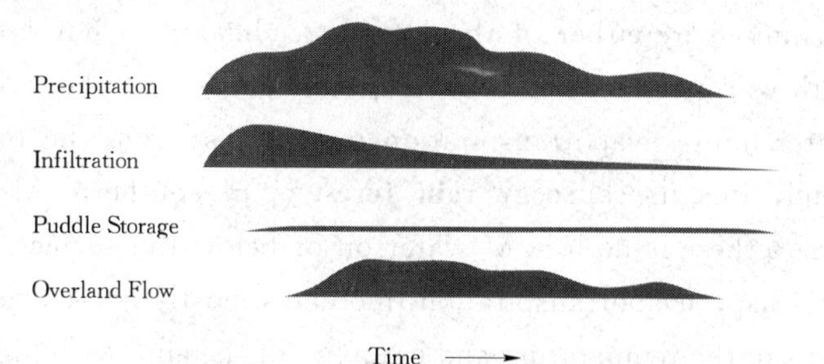

Fig. 1. 6 Schematic showing rates of precipitation, infiltration, puddle storage, and overland flow during one precipitation event

Fig. 1. 7 shows seasonal groundwater level variations observed in a **nonpumping well**[24] in eastern **Maine**[25]. The groundwater level rises dramatically in springtime as rains and snowmelt create a pulse of infiltration and recharge, and then it declines during dry times in late summer and fall.

[25]Maine *n.* 缅因州
[26]screened['skriːnd] *a.* 过筛的

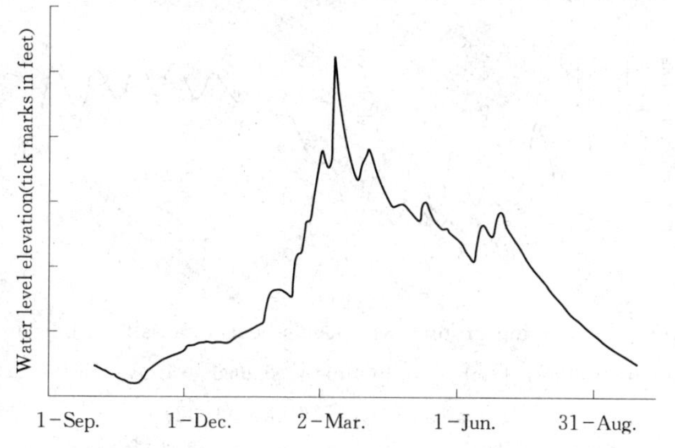

Fig. 1. 7 Ground water levels measured in a well **screened**[26] in a shallow **unconfined**[27] glacial sand and gravel deposit in eastern Maine, 1997 - 1998. From U. S. Geological Survey (Nielsen et al. 1999)

Evapotranspiration[28]

Evapotranspiration refers to the combined processes of direct evaporation at the ground surface, direct evaporation on plant surfaces, and transpiration. Some plants like **beech trees**[29] have shallow root systems to **intercept**[30] infiltrated water before other plants can. Other plants like **oak trees**[31] have deep root systems that can **tap**[32] the saturated zone for a more consistent water supply.

Evapotranspiration rates are governed by several factors, the most important of which are the temperature and

[27]unconfined[ˌʌnkən'faind] *a.* 无限制的，非承压的
[28]evapotranspiration [iˌvæpəʊˌtrænspi'reiʃən] *n.* 土壤水分蒸发蒸腾损失总量
[29]beech trees 山毛榉
[30]intercept[ˌintə'sept]*n.* 遮断，截距，截取 *vt.* 侦听，截取，截听
[31]oak trees 橡树，栎
[32]tap[tæp] *n.* 丝椎，水龙头，分接头 *vi.* 攻丝，轻叩

[33]humidity[hjuːˈmidəti] *n.* 湿度，潮湿，湿气

[34]bone - dry sand dunes 极干的沙丘，砂

[35]soggy rain forest 湿润的雨林，常雨乔木群落

humidity[33] of the air and the availability of water on the surface and in the shallow subsurface. Evapotranspiration can be limited by either of these factors. Imagine a hot desert with warm dry air and **bone - dry sand dunes**[34]. The lack of water limits evapotranspiration in this case. On the other hand, imagine a **soggy rain forest**[35] in southern Alaska where there is no lack of water on or below the surface. In this case, evapotranspiration fluctuates mostly due to variations in the temperature and humidity of the air, with higher evapotranspiration rates on warmer, dryer days.

[36]alfalfa[ælˈʃælʃə]*n.* 苜蓿，紫花苜蓿

Evapotranspiration has daily and seasonal variations. Fig. 1. 8 shows daily fluctuations in groundwater levels beneath an **alfalfa**[36] field. Groundwater levels fell during the day when evapotranspirarion was high and recovered at night when it ceased. When the field was cut, the groundwater level rose due to the lack of transpiration.

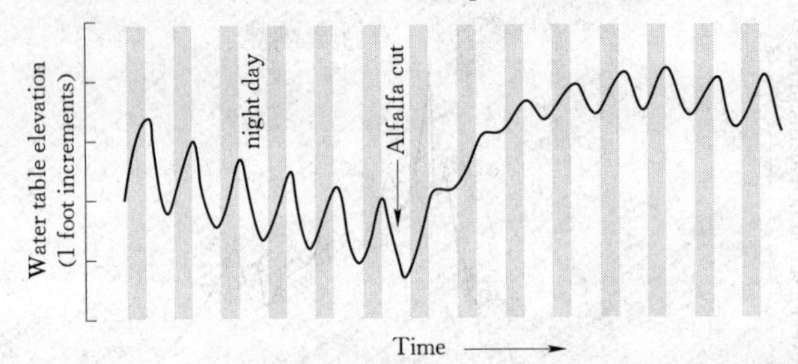

[37]shaded[ˈʃeidid] *a.* 荫蔽的

Fig. 1. 8 Fluctuating groundwater levels beneath an alfalfa field in the Escalante Valley, Utah. Night time is **shaded**[37] gray. From U. S. Geological Survey (White, 1932)

In regions with strong seasonal climate variations, the rate of evapotranspiration is generally lower in winter than in summer because less water can evaporate into cool air than into warm air. Where winters are cold enough for snow and ice, there is very little evapotranspirarion during winter.

[38]plus[plʌs]*a.* 正的 *n.* 正号，加 *prep.* 加 *vt.* 加

As you can imagine, the rate of evapotranspiration cannot be measured directly; that would require somehow measuring the flux of water through all plants **plus**[38] direct evaporation. It is often estimated by developing estimates of other fluxes shown in Fig. 1. 5 and then deducing the estimated evapotranspiration by water balance. For some agricultural crop settings, evapotranspiration has been estimated by carefully studying a small plot of soil: measuring precipitati-

on and changes in **water content**[39] in the soils. This type of study is sometimes conducted using large soil **tanks**[40] called **lysimeters**[41], which contain typical soils and vegetation and are buried in the ground. The mass of the lysimeter and fluxes of water are measured to determine the rate of evapotranspiration.

Thornthwaite (1948) introduced the **theoretical**[42] concept of potential evapotranspiration (PET), which is the amount of evapotranspiration that would occur from the land surface if there were a continuous and unlimited supply of soil moisture. PET is only a **function**[43] of **meteorological factors**[44] such as temperature, humidity, and wind. In a very rainy climate with **soggy soil**[45] the actual evapotranspiration (AET) is essentially equal to PET, but in a dry climate AET is much less than PET. Many crops need to transpire water at rates close to the PET rate. In areas where precipitation is much less than PET, considerable irrigation is required. In the dry high plains of the Midwestern U. S, precipitation during the summer is only a small fraction of PET and irrigation is widespread. In much of the humid eastern U. S., summer precipitation equals or exceeds PET, and many farms survive without irrigation.

[39] water content 含水量，含水率

[40] tank[tæŋk] *n.* 振荡回路，筒，柜

[41] lysimeter[lai'simitə] *n.* 渗水计，测渗计，渗漏测定计

[42] theoretical[ˌθiə'retikl] *a.* 理论的

[43] function['fʌŋkʃn] *n.* 函项，函数，作用

[44] meteorological factors 气象因素

[45] soggy soil 湿润土壤

Groundwater Discharge to Surface Water Bodies

All water that you see flowing in a stream **originates**[46] as precipitation, but the water takes various routes to get there. Some runs directly over the land surface to a channel (overland flow). Some infiltrates a little way and runs horizontally in near-surface soils to a channel (interflow). Some infiltrates deeply to become recharge and then migrates in the saturated zone to discharge back to the surface at a spring, lake, or stream channel. The portion of stream flow that is attributed to this latter path is called **base flow**[47].

[46] originate[ə'ridʒineit] *vt.* 发源，创造，发自 *n.* 起始，起源

If the geologic materials in a stream basin are very permeable, baseflow can be a large part of a stream's discharge. If the materials have low permeability, most precipitation does not infiltrate and baseflow is only a small portion of stream discharge. Baseflow is a fairly steady **component**[48] of a stream's discharge, maintaining low flows dur-

[47] base flow 基流

[48] component[kəm'pəunənt] *n.* 元件，成分，组件 *a.* 组成的

ing periods of **drought**[49]. Flow **contributed**[50] by overland flow of shallow interflow is more transient and occurs during and soon after precipitation events.

[49]drought[draʊt] *n.* 干旱 *a.* 干旱的

[50]contribute *n.* 贡献 *vt.* 有助于 *vi.* 有助于

In humid climates, there is generally discharge from the saturated zone up into surface water; such streams are called gaining steams (see Fig. 1. 10). The top of the saturated zone in the **adjacent**[51] terrain is above the water surface elevation of gaining streams. In losing streams, water discharges in the other direction: from the stream to the subsurface. These situations generally occur in **arid climates**[52], where the depth to saturated zone is great. The base of a losing stream may be above the top of the saturated zone as shown in the right side of Fig. 1. 9 or it may be within the saturated zone.

[51]adjacent[ə'dʒeisənt] *a.* 邻接的，邻近的，邻近（地面）

[52]arid climates 干燥气候

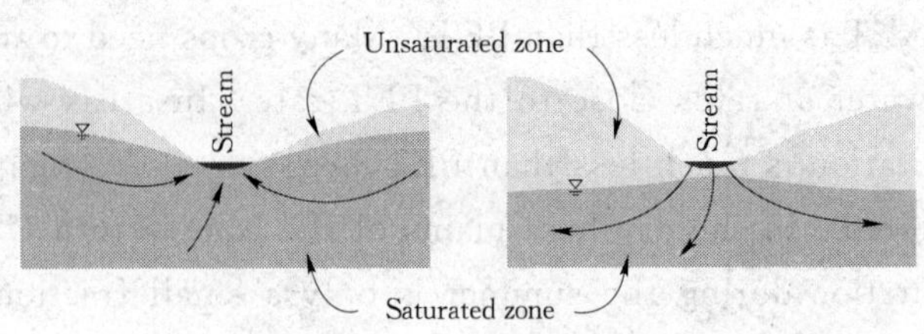

Fig. 1. 9 Cross－sections of a gaining stream (left) and a losing stream (right)

Fig. 1. 10 shows plots of steam discharge vs. time at specific locations on two different streams. This type of plot is known as a stream **hydrograph**[53]. Both streams are in northern **Indiana**[54] and experience similar precipitation patterns. The geology of the two basins differs, however, and causes quite different stream discharge patterns. The low permeability **clayey soils**[55] in Wildcat Creek's basin limit infiltration, recharge, and base flow to a small fraction of precipitation, and most of the steam discharge comes from overland flow (Manning, 1992). On the other hand, the Tippecanoe River basin has sand and gravel soil and most of the stream discharge there is base flow (Manning, 1992). The overland flow that **dominates**[56] the discharge of Wildcat Creek fluctuates greatly in response to precipitation events, while the base flow discharge of the Tippecanoe River **undergoes**[57] mild fluctuations.

[53]hydrograph['haidrəgræf] *n.* 水文曲线，水位图

[54]Indiana[ˌindi'ænə] *n.* 印地安那州

[55]clayey soil 黏质土

[56]dominate['dɒmineit] *vt.* 支配，统治

[57]undergo[ʌndə'gəu] *vt.* 经历

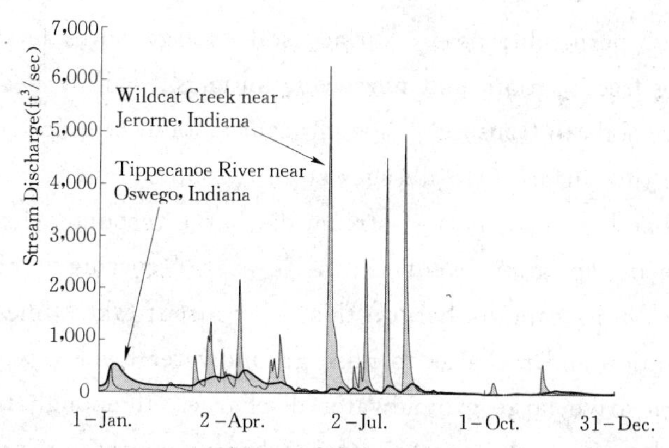

Fig. 1. 10 Hydrographs for two nearby streams in northern Indiana during 1998. The basin upstream of the gage on the Tippecanoe River is 113 mi^2, and the basin upstream of the Wildcat Creek **gage**[58] is 146 mi^2. From U. S. Geological survey (http://www. waterdata. usgs. gov/nwis-w/US/)

[58] gage [geidʒ] *n.* 抵押，规，表

[59]recede[ri'si:d] *vt.* 后退，贬值 *vi.* 后退

[60]portion['pɔ:ʃn] *n.* 部分，分与财产，一部分

Fig. 1. 11 shows a hypothetical steam hydrograph from a single precipitation event. The stream discharge rises during the event and then gradually **recedes**[59] after the event. Fig. 1. 11 illustrates a schematic separation of the total steam discharge Q_s into the base flow portion Q_b and the quickflow **portion**[60] Q_q:

$$Q_s = Q_b + Q_q \tag{1.2}$$

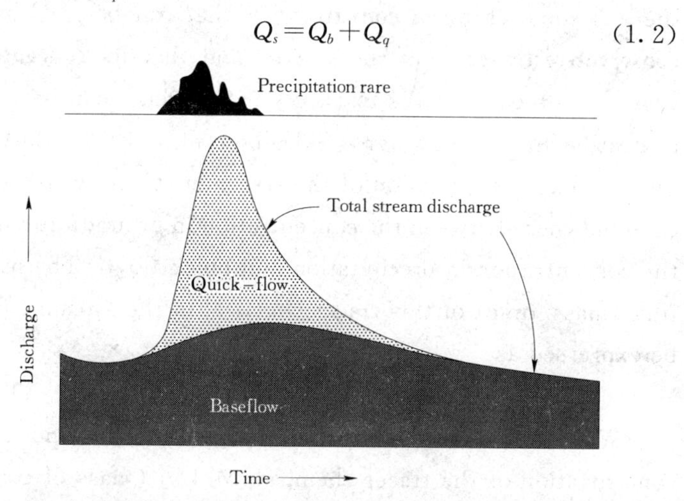

Fig. 1. 11 Hypothetical stream hydrograph during a single precipitation event, showing the contributions of base flow and quick flow to the total stream discharge

Quickflow is the discharge that reaches the stream channel quickly following a precipitation event, and it consists of overland flow plus shallow interflow that moves quickly

[61] rotting tree 腐败，腐朽的树木
[62] burrowing animals 掘穴动物
[63] transmit[træns'mit] vt. 传送，透射，传播
[64] occur[ə'kɜː] vi. 发生；出现
[65] onset['ɒnset] n. 波至，攻击，发动
[66] exponential decay 衰变指数
[67] curve[kɜːv] n. 曲线，弯曲，圆括号 vt. 弯曲
[68] recession[ri'seʃ(ə)n] n. 后退，撤回，不景气
[69] hydrology textbooks 水文学教材
[70] chemical constituent 化学成分，组分，组成成分
[71] conservative tracer 防腐剂示踪物
[72] concentration [kɔns(ə)n'treiʃ(ə)n] n. 集中，浓度，浓缩
[73] markedly['mɑːkidli] ad. 显著地

through permeable near - surface soil. Large pores left by **rotting tree**[61] roots and **burrowing animals**[62] allow near - surface soils to **transmit**[63] a significant amount of quick flow below the surface (Hornberger et al., 1998).

The base flow part of stream discharge responds slowly to the precipitation event, with its peak **occurring**[64] after the peak in steam discharge, this is because it takes time for infiltration and recharge to raise groundwater levels, which in turn drive large groundwater discharge. If enough time passed without the **onset**[65] of another precipitation event, the stream discharge may become entirely base flow. As base flow recedes following a precipitation event it tends to so with an **exponential decay**[66]; this portion of the **curve**[67] is called the base flow **recession**[68].

Empirical, graphical methods for estimating the base flow and quick flow components in a stream hydrograph can be found in most **hydrology textbooks**[69] (see Bras, 1990, for example). Other base flow estimation methods are based on measurements of the chemistry of precipitation, groundwater, and stream water. This technique requires that there is some **chemical constituent**[70] that can be used as a **conservative tracer**[71] in the water, and that its **concentration**[72] in precipitation is **markedly**[73] different than its concentration in groundwater (Sklash et al., 1976; Buttle, 1994). The concentration of the tracer in stream water will be somewhere between the concentration in groundwater and the concentration in precipitation and quick flow. The mass flux (mass/time) of this tracer chemical in the stream could be expressed as

$$Q_s c_s = Q_b c_b + Q_q c_q \tag{1.3}$$

[74] subscript['sʌbskript] n. 下标，添标，索引

Where: Q indicates water discharge[L^3/T], c indicates concentration of the tracer chemical[M/L^3] (mass of tracer per volume of water), and the **subscripts**[74] s, b and q refer to stream, base flow, and quick flow, respectively. It is assumed that c_q equals the concentration of the tracer in the precipitation (and quickflow) and c_b equals the concentration of the tracer in groundwater (and baseflow). Assuming that measurements can be made of Q_s, c_s, c_b, and c_q, the baseflow Q_b and quick flow Q_q discharges can be calculated with Eq. (1.2) and Eq. (1.3).

Pumping

Wells **drilled into**[75] the saturated zone are **pumped**[76], extracting groundwater and transferring it to other reservoirs. Most of the water pumped for irrigation is ultimately transferred to the atmosphere through evaporation and transpiration. Farmers have an economic, **incentive**[77] to pump and irrigate with the minimum discharge that will provide **sufficient transpiration**[78] for the crops. Pumping at higher rates will route more water to evaporation and to deep recharge, both forms of **wastage**[79] in the farmer's view.

In rural homes where pumped water is routed to a **septic system**[80], most of the water returns to the ground as infiltration, although some water evaporates in **shower**[81], laundry machines, outdoor use, etc. Where there are sewer systems, domestic water is collected, treated, and usually discharged to surface water. In some areas like Long Island, New York, pumping of groundwater combined with **municipal**[82] sewer systems amounts to **siphoning**[83] off groundwater and routing it to the sea. As a consequence, groundwater levels fall as the groundwater reservoir shrinks. To limit this problem, **municipalities**[84] in Long Island now direct some of the treated **sewage water**[85] back to the subsurface in infiltration basins and **injection wells**[86], rather than routing it to the sea.

(2309 words)

[75]drilled into 用（机器、工具等）在（某物）上钻孔

[76]pump[pʌmp] n. 泵，心脏，抽运

[77]incentive[in'sentiv]n. 刺激，诱因，动机 a. 激励的

[78]sufficient transpiration 充分蒸发

[79]wastage['weistidʒ] n. 废物，损失，消耗

[80]septic system 腐败性系统

[81]shower['ʃauə(r)]n. 阵雨，淋浴，淋洒器 vt. 浇灌

[82]municipal [mju'nisip(ə)l] a. 市政的，市的；地方自治的

[83]siphon['saifən]n. 虹吸管，无隔菌丝，虹吸 vt. 吮吸

[84]municipality [mju'nisə'pæləti] n. 市区

[85]sewage water 污水

[86]injection well 注入井

Exercises

1. Translate the following sentences into Chinese.

(1) It can infiltrate into the ground to become infiltration, it can flow across the ground surface as overland flow, or it can evaporate from the surface after the precipitation stops.

(2) Infiltration is favored where there is porous and permeable soil or rock, flat topography, and a history of dry conditions.

(3) All water that you see flowing in a stream originates as precipitation, but the water takes various routes to get there.

(4) Most of the water pumped for irrigation is ultimately transferred to the atmosphere through evaporation and transpiration.

2. Translate the following sentences into English.

(1) 非饱和带的水向下运动进入饱和带，这个过程称为补给。

(2) 城市和郊区的发展产生了许多不透水面，如屋顶、道路和混凝土建筑物，这使得下渗损耗从而地表径流增加。

(3) 当降雨速率大于下渗率时，降雨开始在地表填洼。

(4) 蒸散发是指地表的直接蒸发、植物表面的直接蒸发和蒸腾的结合过程。

(5) 河流流量的基流对降水反应缓慢，其峰值发生在蒸汽流量峰值以后。这是因为它需要时间渗透和补给地下水抬高水位，这反过来驱使大量地下水排泄。

(6) 大多数为了灌溉而抽出的水，最终都会通过蒸发和蒸腾作用转移进大气中去。

Learning section

Tab. 1. 6　　Common Abbreviations 5

学位		
B. A.	Bachelor of Arts	文学学士
B. Arch.	Bachelor of Architecture	建筑学士
B. B. A.	Bachelor of Business Administration	工商管理学士
B. Ed.	Bachelor of Education	教育学士
B. Eng., B. E	Bachelor of Engineering	工学学士
B. S.	Bachelor of Science	理学学士
B. M.	Bachelor of Music	音乐学士
LL. B	Bachelor of Law	法学学士
M. A	Master of Arts	文学硕士
M. B. A	Master of Business Administration	工商管理学硕士
M. Div	Master of Divinity	神学硕士
M. E	Master of Engineering	工学硕士
M. Ed	Master of Education	教育学硕士
M. F. A	Master of Fine Arts	艺术硕士
M. L	Master of Law	法学硕士
M. S	Master of Science	理学硕士
Ph. D	Doctor of Philosophy	哲学博士
D. E	Doctor of Engineering	工学博士

Unit 6. Hydrologic Balance

Passage

Hydrologic balance[1] is the basic concept of **conservation of mass**[2] with respect to water fluxes. Take any region in space, and examine the water fluxes into and out of that region. Because water cannot be created or destroyed in that region, hydrologic balance requires

flux in－flux out = rate of change in water stored within (1.4)

The units of each term in this equation are those of discharge[L^3/T]. This is a volume balance, but because water is so **incompressible**[3], it is **essentially**[4] a mass balance as well. Hydrologic balance is useful for estimating unknown fluxes in many different hydrologic systems.

Example 1.1 Consider a reservoir with one inlet stream, one outlet at a dam and a surface area of 2.5km^2. There hasn't been any rain for weeks, and the reservoir level is falling at a rate of 3.0mm/day, and the average evaporation rate from the reservoir surface is 1.2mm/day, the inlet discharge is 10,000m^3/day. And the outlet discharge is 16,000m^3/day. Assuming that the only other important fluxes are the groundwater discharges in and out of the reservoir, what is the total net rate of groundwater discharge into the reservoir?

In this case, the reservoir is the region for which a balance is constructed. Fluxes into this region include the inlet stream flow (I) and the net groundwater discharge (G). Fluxes out of this region include the outlet stream flow (O) and evaporation (E) from the surface. Hydrologic balance in this case requires

$$I+G-O-E=\frac{dV}{dt}$$

Where: dV/dt is the change in reservoir volume per time. Calculate the rate of change in reservoir volume as

[1]hydrologic balance 水文均衡
[2] conservation of mass 质量守恒
[3]incompressible [inkəm'presib(ə)l] *a.* 非压缩的，不可压缩的，不能压缩的
[4]essentially[i'senʃəli] *ad.* 本质上

$$\frac{dV}{dt} = -0.003\ \frac{m}{day} \times 2.5km^2 \times \left(1000\ \frac{m}{km}\right)^2 = -7500\ \frac{m^3}{day}$$

Similarly, the rate of evaporative loss is

$$E = 0.0012\ \frac{m}{day} \times 2.5km^2 \times \left(1000\ \frac{m}{km}\right)^2 = 3000\ \frac{m^3}{day}$$

Solving the first equation for G **yields**[5] a net groundwater discharge of $1,500m^3/day$.

Fluxes in and out of the saturated zone of an aquifer in a stream basin are illustrated in Fig. 1.12. Using symbols defined in the Figure's **caption**[6], the general equation for hydrologic balance in this **piece**[7] of aquifer is as follows:

$$R + G_i - G_o - G_s - ET_d - Q_w = \frac{dV}{dt} \tag{1.5}$$

Where: dV/dt is rate of change in the volume of water stored in the region.

If, over a long time span, there is an approximate steady-state balance where flow in equals flow out, then the transient term disappears and the balance equation becomes

$$R + G_i - G_o - G_s - ET_d - Q_w = 0 \tag{1.6}$$

Imagine that the basin has been **operating**[8] in a rough steady state for many years without any **pumping wells**[9]. At this time, the above equation with $Q_w = 0$ describes the balance layer a well is installed and begins pumping at a steady rate $Q_w > 0$. Immediately, the system is thrown into imbalance; Eq. (1.4) applies and volume stored in the aquifer declines ($dV/dt < 0$). In fact, at the start of pumping, all the water pumped comes from a corresponding decline in the volume of water stored ($Q_w = -dV/dt$). The decline in volume stored causes a declining water table. If the well discharge is held constant for a long time, a new long-term equilibrium will be established, the water table will stabilize, and Eq. (1.5) will apply once again, this time with $Q_w > 0$. To achieve this new long-term balance, other fluxes must adjust: R and G_i may increase, while G_o, G_s and ET_d may

[5]yield[ji:ld]*n*. 产量，产额

[6]caption['kæpʃn]*n*. 标题，说明文字

[7]piece[pi:s] *n*. 块，片，段；部分，部件；文章，音乐作品 *vt*. 修补；连接，接上

[8]operating['ɔpəreitiŋ] *a*. 运行的 *n*. 操作，运转

[9]pumping wells 抽油井

decrease.

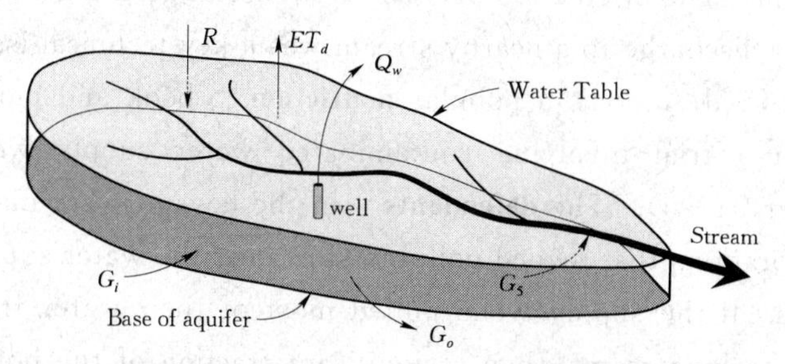

Fig. 1. 12 Water fluxes in and out of the saturated zone of an aquifer under a stream basin R is recharge, G_i and G_o are groundwater inflows and outflows through the lateral boundaries and bottom of the aquifer, G_s is groundwater discharge to streams, ET_d is deep evapotranspiration extracted form the saturated zone, and Q_w is well discharge

Increasing the discharge of pumping wells in a groundwater basin always has some long - term effects on other fluxes and/or the volume stored in the basin. In many cases, the rate of pumping is so high that a new steady state cannot develop; the flows out of the system are greater than the flows in. In such cases, the pumping could be viewed as "mining," simply pulling water out of storage. This has been the case in the High Plains Aquifer in the south - central U. S. , where there is wide spread irrigation and a dry climate. The water table has dropped more than 30m in parts of this aquifer in **Texas**[10], where the average rate of pumpage far exceeds the average recharge rate, which is estimated to be less than 5mm/year (Gutentag et al. ,1984). By 1980, pumping had removed about 140km^3 of stored water from the aquifer in Texas.

[10] Texas['teksəs] *n.* 德克萨斯州

Long - term changes caused by pumping can affect large areas, far beyond the well owner's property lines. Groundwater basins often **transgress**[11] property lines and political boundaries, so most countries have regulations governing groundwater use. Regulation of well discharges and groundwater resources is complex because of interaction between different aquifers, surface water bodies, and wells.

[11] transgress[træns'gres] *vt.* 违犯，违反

Often pumping wells located near streams cause a noticeable reduction in groundwater discharge to the stream. In other words, the wells steal some base flow from the stream. The interaction between well discharge and groundwater discharge to a nearby stream was a key technical issue in A Civil Action, a popular **nonfiction**[12] book and movie about a trial involving contaminated water supply wells (Harr, 1995). The **defendants**[13] in the lawsuit were major corporations that owned polluted sites near the water supply wells. If the supply wells pulled most of their water from the nearby stream, then a significant fraction of the pollutions in the well water could have come from distant sites upstream along the stream, and the owners of local sites would have been less **culpable**[14].

(912 words)

[12]nonfiction[nɔn'fikʃən] *n.* 非小说文学

[13]defendant[di'fendənt] *n.* 被告

[14]culpable['kʌlpəbl] *a.* 有罪的，应受处罚的

Exercises

1. Translate the following sentences into Chinese.

(1) Hydrologic balance is the basic concept of conservation of mass with respect to water fluxes.

(2) Increasing the discharge of pumping wells in a groundwater basin always has some long-term effects on other fluxes and/or the volume stored in the basin.

(3) This has been the case in the High Plains Aquifer in the south-central U. S., where there is wide spread irrigation and a dry climate.

(4) Groundwater basins often transgress property lines and political boundaries, so most countries have regulations governing groundwater use.

(5) If the supply wells pulled most of their water from the nearby stream, then a significant fraction of the pollutions in the well water could have come from distant sites upstream along the stream, and the owners of local sites would have been less culpable.

2. Translate the following sentences into English.

(1) 在许多不同的水文系统中，水文平衡对估计未知的流量十分有用。

(2) 区域流入量包括入流量（*I*）和净地下水排泄量（*G*）。区域流出量包括出流量（*O*）和地面蒸发量（*E*）。

(3) 地下水流域内，抽水井排泄量的增加对流域内储存的其他通量与体积有长期影响。

(4) 在德克萨斯州的部分含水层，水位已经下降了超过 30m，泵送平均率远远超过了平均充电泵的速度，估计不到 5mm/年。

(5) 关于水井开采和地下水资源的法规十分复杂，因为不同的含水层、地表水、水井之间会有相互交换。

Learning section

Tab. 1.7　　Common Abbreviations 6

世界主要组织		
UNESCO	United Nations Educational, Scientific and Cultural Organization	联合国教育科学文化组织
NASA	National Aeronautics and Space Administration	美国航天太空总署
WHO	World Health Organization	世界卫生组织
FBI	Federal Bureau of Investigation	联邦调查局
CIA	Central Intelligence Agency	中央情报局
USDA	United States Department of Agriculture	美国农业部
USGS	United States Geological Survey	美国地质勘探局
MWR	Ministry of Water Resources	水利部

Unit 7. School of Water Conservancy[1] and Electric-power

Passage

Four types of rooms can be classified in the building of School of Water Conservancy and Electric Power of Heilongjiang University: fundamental facility rooms, special classrooms, laboratories and administrative offices, all of which will be introduced in detail in the following passage.

Fundamental facility rooms provide basic facilities and services for the staff and students, which are generally set up in almost all kinds of buildings. Room of Reception and Delivery and **Estate**[2] Management are often located on the first floor while Lady's and Gentleman's room are located on each floor. Particularly, a Staff Club is built on the top floor with sports facilities and space.

Special classrooms are classrooms with special functions, such as Drawing Room for the course of **Engineering Drafting**[3], Digital Designing and Mapping Room for the course of AutoCAD, Academic Lecturing Hall for the purpose of holding lectures and activities, **Postgraduates**[4] classroom for the purpose of giving classes and seminars by postgraduates only.

Laboratories under the three different departments and one institute can be divided into 4 specialized laboratories: Labs of **Hydraulic and Hydropower Engineering**[5], Labs of **Hydrology**[6] and Water Resources Engineering, Labs of **Agricultural**[7] Water Conservancy Engineering and Labs of **Institute of Groundwater in Cold Region**[8]. 9 **Hydraulics**[9] laboratories and 10 **Geotechnical**[10] Engineering laboratories are under the branch of Hydraulic and Hydropower Engineering. Under the branch of Hydrology and Water Resources Engineering, there are 2 **Meteorology**[11] laboratories, 1 **Hydrochemical**[12] Analysis Instrument Room, 1 **Balance**[13] Room, 1 **Preheating**[14] Room, 1 Storage of Chemical Drug Room, 1 **Glassware**[15] Instrument Room and 2 Students Labo-

[1]Water Conservancy 水利
[2]estate[i'steit] *n.* 土地，财产
[3]engineering drafting 工程制图
[4]postgraduate [ˌpəust'grædʒuət] *n.* & *a.* 研究生，研究生的
[5] Hydraulic and Hydropower Engineering 水利水电工程
[6]hydrology[hai'drɒlədʒi] *n.* 水文学. hydro-水文的前缀
[7] agricultural [ˌægri'kʌltʃərəl] *a.* 农业的
[8] Institute of Groundwater in Cold Region 寒区地下水研究所
[9]hydraulics[hai'drɔːliks] *n.* 水力学
[10]geotechnical[dʒiːəu'teknikəl] *a.* 土力学的
[11]meteorology[ˌmiːtiə'rɒlədʒi] *n.* 气象学
[12]hydrochemical[haidrəu'kemikl] *a.* 水化学的
[13]balance['bæləns]*n.* 天平
[14]preheat[ˌpriː'hiːt] *vt.* 预热
[15]glassware['glɑːswɛə] *n.* 玻璃器具类

ratories of Water Environment. Labs of Agricultural Water Conservancy Engineering consists of 2 laboratories: Agricultural **Irrigation**[16] Lab and **Pump and Pumping Station**[17] Lab. Institute of Cold Region Groundwater is constituted by 1 Director Office, 2 laboratories and 2 research rooms, which respectively are Lab of **Ground Ice**[18], Lab of Groundwater **Seepage**[19] **Simulation**[20] in Cold Region, Research Room of **Geothermal**[21] Water and **Mineral**[22] Water in Cold Region and Research Room of Water Resources in Russian **Far East**[23].

There are three types of administrative offices: management offices, student and teaching affairs offices and department offices. Details please refer to the Tab. 1. 8.

Tab. 1. 8 Classifications of Administrative Offices of School of Water Conservancy and Electric Power, HLJU

Classifications	Names of Offices	Room No.
Management Office	**Dean**[24]	401
	CPC[25] Secretary	402
	Deputy[26] Dean	403
	Vice[27] - secretary, CPC	404
	Office of Dean Assistant	407
	CPC Assistant	408
	Administrative Office	409
	Asset[28] Management Office	410
Student and Teaching Affairs Office	Teaching inspection and direction	406
	Archive[29]	501
	Office of Postgraduates	413
	Office of Teaching Affairs	411
	Office of Student Affairs	412
Department Office	Office of Discipline Development	424
	Head of Department	405
	Department of Hydraulic and Hydropower Engineering	419
	Department of Hydrology and Water Resources Engineering	421
	Department of Agricultural Water Conservancy Engineering	423

(415 words)

[16] irrigation[ˌiriˈgeiʃn] *n.* 灌溉
[17] Pump and Pumping Station 水泵与泵站
[18] ground ice 地下冰
[19] seepage[ˈsiːpidʒ] *n.* 渗流
[20] simulation[ˌsimjuˈleiʃn] *n.* 模拟，模仿
[21] geothermal[ˌdʒi(ː)əuˈθəməl] *a.* 地热的，地温的
[22] mineral[ˈminərəl] *n.* 矿物
[23] Far East 远东
[24] dean[diːn] *n.* 院长
[25] CPC: Communist Party of China 中国共产党
[26] deputy[ˈdepjuti] *n.* 副手代理人，议员
[27] vice[vais] *a.* 副的
[28] asset[ˈæset] *n.* 有价值的人或物，资产，财产
[29] archive[ˈɑːkaiv] *n.* 档案，档案馆

Exercises

1. Build sentences with following words.

(1) room distribution, academic building, purpose

(2) special classrooms, engineering drafting, AutoCAD, academic lecturing hall

(3) specialized laboratories, Hydraulic and Hydropower Engineering, Hydrology and Water Resources Engineering, Agricultural Water Conservancy Engineering

2. Translate the following phrases into Chinese.

Room of Reception and Delivery	
Digital Designing and Mapping Room	
Hydraulics Lab	
Geotechnical Engineering Lab	
Meteorology Lab	
Hydrochemical Analysis Instrument Room	
Storage of Chemical Drug Room	
Agricultural Irrigation Lab	
Pump and Pumping Station Lab	
Lab of Ground Ice	
Lab of Groundwater Seepage Simulation in Cold Region	
Research Room of Geothermal Water and Mineral Water in Cold Region	
Research Room of Water Resources in Russian Far East	

3. Translate the following sentences into English.

(1) 黑龙江大学水利电力学院的房间按功能可划分为四种类型：①基础功能房间、②专门教室、③实验室、④行政办公室。

(2) 基础功能房间共有四间，分别为：①收发室、②物业管理室、③男女卫生间、④职工之家。

(3) 水电楼共有四类专门教室，包括：①制图室、②数字化设计室、③学术报告厅、④研究生教室。

(4) 水电学院实验室可分为 4 类：①水工实验室、②水文实验室、③农水实验室、④寒区地下水研究所实验室。

(5) 水电学院大楼 4 层设有院长室、书记室、学工办、教务办，以及水文与水资源、农业水利工程及水利水电工程三个系办公室。

Learning section

Tab. 1. 9　　Expression of Mathematics

Sets (集合)	
$x \in A$	x belongs to A/x is an element (or a member) of A
$x \notin A$	x does not belong to A/x is not an element (or a member) of A
$A \subset B$	A is contained in B/A is a subset of B
$A \supset B$	A contains B/B is a subset of A

Unit 8. Curriculum for Hydrology and Water Resources Engineering

Passage

Tab. 1. 10 Names and Credits for Compulsive Courses

No.	Courses	Credits
1	Advanced Mathematics Ⅰ Ⅱ	10
2	Linear **Algebra**[1] Ⅲ	2
3	College Physics	4
4	Introduction of Profession	1
5	Engineering Drafting	2
6	Physical **Geography**[2]	2
7	Engineering Survey	3
8	**Climatology**[3] and **Meteorology**[4]	2. 5
9	Principles of Hydrology	2. 5
10	**Hydraulics**[5]	4
11	Probability Theory & Mathematical **Statistics**[6]	3
12	**General Hydrogeology**[7]	2
13	Water Quality Testing and Hydrochemical Analysis	3
14	**Hydrometry**[8]	3
15	**Groundwater Dynamics**[9]	2. 5
16	**Hydrological Statistics**[10]	2. 5
17	Water Environmental Assessment and Protection	2. 5
18	Hydrological Analysis and Calculation	2. 5
19	Hydrology Forecasting	2. 5
20	Water Resources System Analysis	2
21	Hydraulic Calculation and Water Resources Planning	2. 5
22	Course Exercise of Water Environmental Assessment and Protection	1
23	**Practicum**[11]	1
24	Practice of Engineering Survey	2
25	Experiments in College Physics	2
26	Practice of Physical Geography	1
27	Practice of Hydrology and Meteorology	1

[1]algebra['ældʒibrə]*n.* 代数学，代数

[2]geography[dʒi'ɒgrəfi] *n.* 地理（学）

[3]climatology[ˌklaimə'tɒlədʒi] *n.* 气候学

[4]meteorology[ˌmiːtiə'rɒlədʒi] *n.* 气象学

[5]hydraulics[hai'drɔːliks] *n.* 水力学

[6]statistics[stə'tistiks] *n.* 统计学

[7]general hydrogeology 水文地质学基础

[8]hydrometry[hai'drɒmitri] *n.* 水文测验

[9]groundwater dynamics 地下水动力学

[10]hydrological statistics 水文统计

[11]practicum['præktikəm] *n.* 实习科目，实习课

续表

No.	Courses	Credits
28	Hydrometry Internship	2
29	Course Exercise of Hydrological Analysis and Calculation	1
30	Course Exercise of Hydrology Forecasting	1
31	Course Exercise of Hydraulic Calculation and Water Resources Planning	1
32	Groundwater Comprehensive Practice	1
33	Undergraduate Practicum & Graduation Design/**Dissertation**[12]	10

[12] dissertation [ˌdisəˈteiʃn] *n.* 学位论文

Tab. 1. 11 Names and Credits for Optional Courses

No.	Courses	Credits
1	Principles and Applications of Database	2
2	Electro－technics and Electrical Equipments	2
3	Course of C Program Design	3
4	**Stochastic**[13] Hydrology	2
5	Introduction of Hydraulic and Hydro－Power Engineering	2
6	Engineering Mechanics	3
7	AutoCAD	2
8	Computational Method	2
9	Hydrologic Model of Basin	2
10	Groundwater Development and Utilization	2
11	Groundwater Mathematical Model	2
12	Entrepreneur Management	2
13	Introduction of Natural Resources	2
14	3S Technology	2
15	Environmental **Ecology**[14]	2
16	River Dynamics	2
17	Water Resources Assessment and Management	2
18	Urban Hydrology	2
19	Computer Aided Design	2
20	Literature Search and Application	2
21	**Cold Region**[15] Water Problems Seminar	1
22	Water Disasters Prevention	1
23	Computer Networks	2
24	Instant Flood Forecasting System	1
25	Reservoir Operation	2
26	Professional English	2
27	**Hydrography**[16] and History of Water Conservancy	1

[13] stochastic [stəˈkæstik] *a.* 随机的

[14] ecology [iˈkɒlədʒi] *n.* 生态，生态学

[15] Cold Region 寒区

[16] hydrography [haiˈdrɒgrəfi] *n.* 水文地理学

Exercises

1. Build sentences with the following words.

(1) curriculum, hydrology, water resources engineering

(2) credits, compulsive courses, optional courses

(3) experiments, practice, graduation design/dissertation

2. Translate the following sentences into English.

(1) 水文与水资源工程的核心课程有：自然地理学、水文学原理、气象与气候学、水力学、工程力学、水利水电工程概论、水资源系统分析、地下水动力学、水环境评价与保护等。

(2) 水文与水资源工程专业课程包括必修课和选修课，专业英语课为选修课之一。

(3) 随机水文学是水文与水资源专业的选修课之一，共2个学分。

Learning section

Tab. 1. 12　　Expression of Mathematics 2

Real numbers（实数运算）	
$x+y$	x plus y
$x-y$	x minus y
$x\pm1$	x plus or minus one
$x\times y$	xy/x multiplied by y
$x\div y$	x devided by y
$(x-y)(x+y)$	x minus y, x plus y
$x=5$	x equals 5/x is equal to 5
$x\neq5$	x (is) not equal to 5
$x>y$	x is greater than y
$x\geqslant y$	x is greater than or equal to y
$x<y$	x is less than y
$x\leqslant y$	x is less than or equal to y
$0<x<1$	zero is less than x is less than 1
$0\leqslant x\leqslant1$	zero is less than or equal to x is less than or equal to 1
$\lvert x\rvert$	mod x/modulus x
x^2	x squared/x (raised) to the power 2
x^3	x cubed
x^n	x to the nth/x to the power n
x^{-n}	x to the (power) minus n
$\sqrt{x}$	(square) root x / the square root of x
$\sqrt[3]{x}$	cube root (of) x
$\sqrt[4]{x}$	fourth root (of) x
$\sqrt[n]{x}$	nth root (of) x

Unit 9. Disciplines Distribution of Hydrology and Water Resources Engineering in China

Passage

[1]institute['institjuːt] *n.* 协会，学会，学院，研究院
[2]geoscience[ˌdʒiːəʊ'saiəns] *n.* 地球科学

Tab. 1. 13 Doctoral degree programs of water resources (16 in total)

1	Tsinghua University
2	Dalian University of Technology
3	Hohai University
4	Wuhan University
5	Sichuan University
6	Xi'an University of Technology
7	Tianjin University
8	China **Institute**[1] of Water Resources and Hydropower Research
9	Zhengzhou University
10	Nanjing Hydraulic Research Institute
11	Jilin University
12	Nanjing University
13	China Agricultural University
14	China University of **Geosciences**[2]
15	Chang'an University
16	Northwest Agriculture & Forest University

Tab. 1. 14 Master degree programs of water resources (48 in total)

Province	No.	Affiliation
Anhui	1	HeFei University of Technology
Beijing	5	Beijing Normal University
		Tsinghua University
		Capital Normal University
		China Agricultural University
		China Institute of Water Resources and Hydropower Research
Gansu	2	Lanzhou University
		Lanzhou Jiaotong University
Guangdong	1	Sun Yat-Sen UNIVERSITY

续表

Province	No.	Affiliation
Guangxi	1	Guilin University of Technology
Heilongjiang	1	Northeast Agricultural University
Hebei	1	North China Electric Power University
Henan	3	Henan **Polytechnic**[3] University
		North China University of Water Resources and Electric Power
		Zhengzhou University
Hubei	4	Changjiang River Scientific Research Institute
		Huazhong University of Science and Technology
		Wuhan University
		China University of Geosciences
Hunan	2	Changsha University of Science and Technology
		Hunan Normal University
Jilin	1	Jilin University
Jiangsu	5	Hohai University
		Nanjing University
		Nanjing Hydraulic Research Institute
		Yangzhou University
		China University of **Mining**[4] and Technology
Jiangxi	1	East China of Institute and Technology
Liaoning	2	Dalian University of Technology
		Liaoning Normal University
Inner Mongolia	1	Inner Mongolia Agricultural University
Ningxia	1	Ningxia University
Shandong	5	Jinan University
		Shandong University
		Shandong University of Science and Technology
		Shandong Agricultural University
		Ocean University of China
Shanxi	1	Taiyuan University of Technology
Shaanxi	3	Chang'an University
		Xi'an University of Technology
		Northwest Agriculture and Forest University

[3]polytechnic[ˌpɒliˈteknik]
n. 理工学院
[4]mining[ˈmainiŋ]
n. 采矿（业）

续表

Province	No.	Affiliation
Shanghai	1	Tongji University
Sichuan	1	Sichuan University
Tianjin	1	Tianjin University
Xinjiang	1	Xinjiang Agricultural University
Yunnan	1	Kunming University of Science and Technology
Zhejiang	1	Zhejiang University
Chongqing	1	Chongqing Jiaotong University

Exercises

1. Build sentences with following words.

(1) doctoral degree program, master degree program, bachelor degree program

(2) hydraulics, hydrology, hydroelectricity

(3) water resources, affiliation, province

2. Translate the following sentences into English.

(1) 中国可分为6部分：东北、华北、华东、中南、西南、西北。新疆维吾尔自治区和宁夏回族自治区在西北地区。

(2) 中国设有水文与水资源专业博士点的单位有16家，其中在东北的2家为大连理工大学和吉林大学。

(3) 中国设有水文与水资源专业硕士点的单位有48家，其中黑龙江有1家即东北农业大学。

Learning section

Tab. 1. 15　　Expression of Mathematics 3

Functions（函数）	
$f(x)$	fx / f of x / the function f of x
$\lim_{x\to 0} f(x)$	the limit as x approaches zero
$\lim_{x\to 0^+} f(x)$	the limit as x approaches zero from above
$\lim_{x\to 0^-} f(x)$	the limit as x approaches zero from below
$\lg_x y$	lg y to the base x/lg to the base x of y
$\ln_y$	lg y to the base e/lg to the base e of y/natural lg (of) y

Part 2

Figure 1. Cross Section of a Gravity Dam

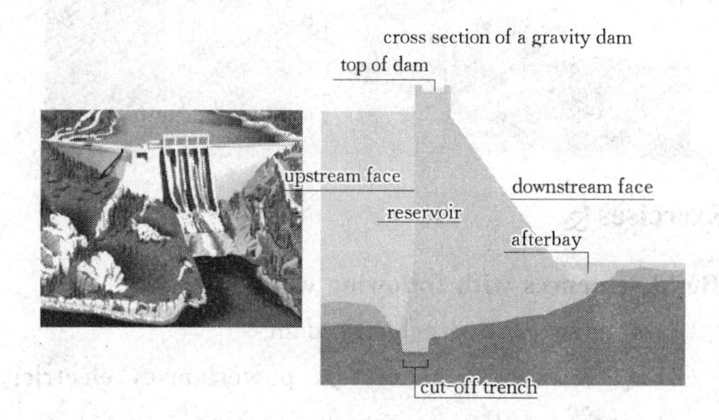

[1]gravity dam 重力坝

[2]cross section 横截面

[3]upstream[ˌʌpˈstriːm]
ad. & *a.* 向上游地（的），逆流地（的）

[4]downstream[ˌdaunˈstriːm]
ad. 在下游，顺流地

[5]reservoir[ˈrezəvwɑː] *n.* 水库，蓄水体

[6]afterbay[ˈɑːftəbei] *n.* 尾水池

[7]trench[trentʃ] *n.* 深沟，地沟；战壕

[8]cut－off trench 齿墙

Exercises

1. Build sentences with following words.

(1) gravity dam

(2) upstream，downstream

(3) afterbay，cut－off trench

2. Translate the following sentences into English.

(1) 齿墙、尾水池、坝顶都是重力坝的重要组成部分。

(2) 坝体上游壅水形成水库，水位高于下游。

3. Try to describe the composition of a gravity dam in English.

Learning section

All rivers run into sea.
海纳百川。
A light heart lives long.
静以修身。

Figure 2. Hydroelectric Power Generation

[1]hydroelectric[ˈhaidrəuiˈlektrik]
a. 水力发电的
[2]intake[ˈinteik] n. 吸入，纳入，（液体等）进入口，进水口
[3]outlet[ˈautlet] n. 出口，出路，出水口
[4]sluice gate 泄水闸门
[5]penstock[ˈpenstɔk]
n. 水道，水渠，压力水管，水阀门
[6]powerhouse[ˈpauəˌhaus] n. 发电厂房
[7]generator[ˈdʒenəreitə] n. 发电机，发生器
[8]turbine[ˈtəːbin,ˈtəːbain] n. 涡轮机
[9]power transmission cable 电力传输电缆
[10]transformer[trænsˈfɔːmə]
n. 变压器
[11]silt[silt] n. 淤泥

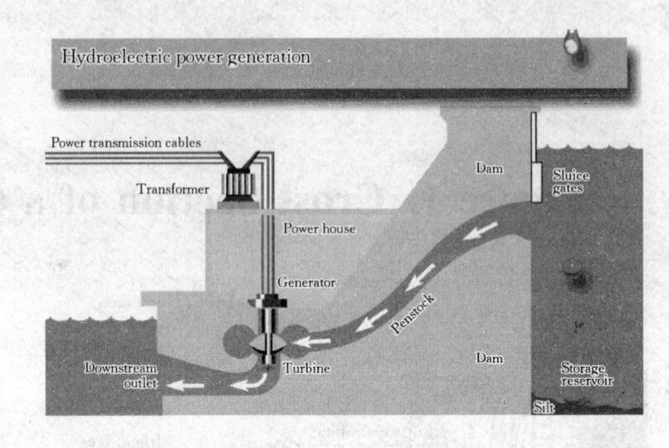

Exercises

1. Build sentences with following words.

(1) structure，hydroelectric dam

(2) water energy，generator，powerhouse，electricity

(3) intake，outlet，penstock，turbine

(4) sluice gate

2. Translate the following sentences into English.

(1) 上游水库中的水通过坝体的压力管道输送至涡轮机发电，之后从出口排出到下游。

(2) 水电站发的电可通过电力传输电缆传给用户。

3. Try to describe the process of hydro - electricity generation in English.

Learning section

There is no smoke without fire.

无风不起浪。

Rome is not built in a day.

冰冻三尺，非一日之寒。

Figure 3. Generator

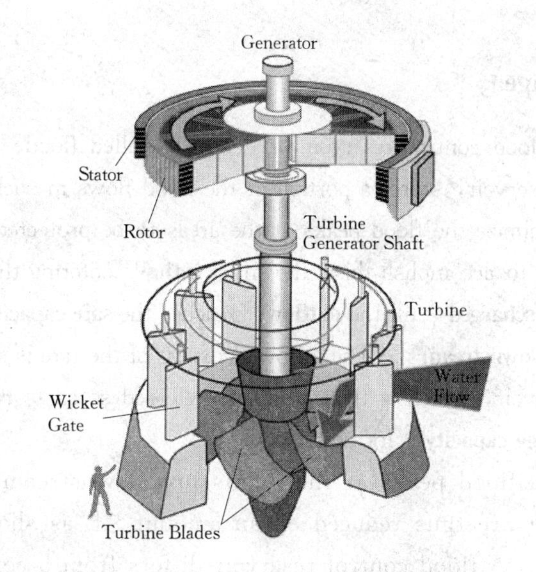

[1] generator['dʒenəreitə] n. 发电机，发生器

[2] shaft[ʃɑ:ft] n. 轴

[3] generator shaft 发电机轴

[4] stator ['steitə] n. 定子，固定片

[5] rotor['rəutə] n. 转子

[6] turbine['tə:bin¦'tə:bain] n. 涡轮机

[7] blade[bleid] n. 刀刃，叶片

[8] turbine blade 涡轮叶片

[9] wicket['wikit] n. 三柱门

[10] wicket gate 导叶

Exercises

1. Build sentences with following words.

(1) generator, powerhouse

(2) stator, rotor

(3) generator shaft, turbine blade

2. Translate the following sentences into English.

(1) 发电机由涡轮叶片、导叶、发电机轴、转子和定子组成。

(2) 在发电机中，水推动涡轮叶片，带动导叶发电。

3. Try to describe the structure of a generator in English.

Learning section

Still water run deep. 大智如愚。

Pour water into a sieve. 竹篮子打水一场空。

Unit 1. Flood Control Reservoir

Passage

A flood control reservoir or generally called flood - **mitigation**[1] reservoir, stores a portion of the flood flows in such a way as to minimize the flood peaks at the areas to be protected downstream. To accomplish this, the entire **inflow**[2] entering the reservoir is **discharged**[3] till the **outflow**[4] reaches the **safe capacity of the channel downstream**[5]. The inflow in excess of the rate is stored in the reservoir, which is then gradually released so as to **recover**[6] the **storage capacity**[7] for next flood.

The flood peaks at the points just downstream of the reservoir are thus reduced by an amount AB as shown in Fig. 2. 1. A flood control reservoir differs from a conservation reservoir only in its need for a large **sluiceway**[8] capacity to permit rapid drawdown before or after a flood.

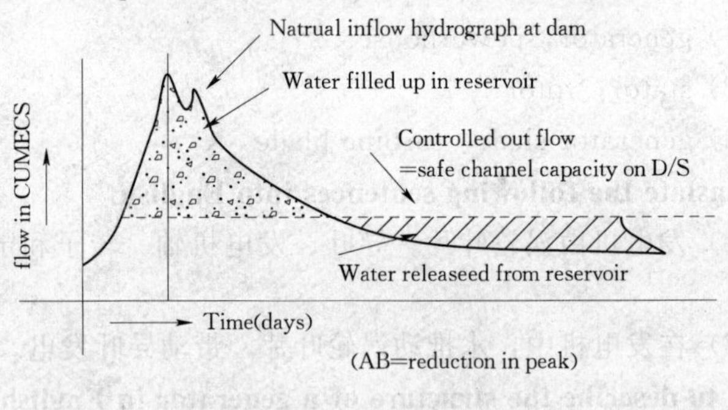

Fig. 2. 1 The principle of flood control reservoirs

Types of flood control reservoirs. There are two basic types of flood - mitigation reservoirs: ①Storage Reservoir or **Detention basins**[9], ②**Retarding basins**[10] or retarding reservoirs.

A reservoir with gates and **valves**[11] **installation**[12] at the **spillway**[13] and at the sluice outlets is known as a storage - reservoir, while on the other hand, a reservoir with fixed **ungated**[14] outlets is known as a retarding basin.

[1]mitigation[ˌmitiˈgeiʃən] *n.* 缓解，减轻，平静 flood - mitigation reservoir 减洪水库，滞洪水库

[2]inflow[ˈinfləu] *n.* 进水量，流入，入流

[3]discharge[disˈtʃɑːdʒ] *n.* 流出，泄流

[4]outflow[ˈɑutfləu] *n.* 流出量；放水，出流

[5] safe capacity of the channel downstream 下游河道安全泄水量

[6]recover[riˈkʌvə] *vt.* 恢复；重新获得，找回

[7] storage capacity 蓄水库容，库容

[8]sluiceway[ˈsluːsˌwei] *n.* 泄洪道，分洪道

[9]detention basin 蓄洪区，拦洪区

[10]retarding basin 滞洪区

[11]valve[vælv] *n.* 阀，活门；(心脏的) 瓣膜；真空管

[12]installation[ˌinstəˈleiʃən] *n.* 安装，设置；就职；装置，设备

[13]spillway[ˈspilwei] *n.* 溢洪道，泄洪道

[14]ungated[ʌnˈgeitid] *a.* 无门的；无闸门的

Functioning and advantages of a retarding basin:

A retarding basin is usually provided with an uncontrolled spillway and an uncontrolled **orifice**[15] type sluiceway. The automatic regulation of outflow depending upon the availability of water, takes place from such a reservoir. The maximum discharging capacity of such a reservoir should be equal to the maximum safe carrying capacity of the channel downstream. As flood occurs, the reservoir gets filled and discharges through sluiceways. As the reservoir elevation increases, outflow discharge increases. The water level goes on rising until the flood has subsided and the inflow becomes equal to or less than the outflow. After this, water gets automatically withdrawn from the reservoir until the stored water is completely discharged. The advantages of retarding basin over a gate controlled detention basin are: ① Cost of gate installations is saved. ② There are no gates and hence, the possibility of human error and **negligence**[16] in their operation is **eliminated**[17]. ③ Since such a reservoir is not always filled, much of land below the maximum reservoir level will be **submerged**[18] only **temporarily**[19] and occasionally and can be successfully used for agriculture, although no **permanent**[20] habitation can be allowed on this land.

Functioning and advantages of a storage reservoir:

A storage reservoir with gated spillway and gated sluiceway, provides more **flexibility**[21] of operation, and thus gives us better control and increased usefulness of the reservoir. Storage reservoirs are, therefore, preferred on large rivers which require better control, while retarding basins are preferred on small rivers. In storage reservoirs, the **flood crest**[22] downstream can be better controlled and regulated properly so as not to cause their **coincidence**[23]. This is the biggest advantage of such a reservoir and **outweighs**[24] its disadvantages of being **costly**[25] and involving risk of human error in installation and operation of gates.

(672 words)

[15] orifice [ˈɔrifis] *n.* 孔口，管口

[16] negligence [ˈneglidʒəns] *n.* 疏忽，粗心大意

[17] eliminate [iˈlimineit] *vt.* 消（排，清）除

[18] submerged [səbˈməːdʒd] *a.* 在水中的，淹没的

[19] temporarily [ˈtempθrərili] *ad.* 暂时地

[20] permanent [ˈpəːmənənt] *a.* 永久(性)的，固定的；永恒的；长久的

[21] flexibility [ˌfleksiˈbiliti] *n.* 柔韧性；机动性，灵活性，易曲性；适应性，弹性

[22] flood crest 洪峰，同 flood peak

[23] coincidence [kəuˈinsidəns] *n.* 符合；一致；同时发生

[24] outweigh [autˈwei] *vt.* 重于，比……重要

[25] costly [ˈkɔstli] *a.* 昂贵的，代价高的，花钱多的；引起困难的；造成损失的

Exercises

1. Translate the following sentences into English.

(1) 防洪水库有两种基本类型：①蓄洪水库或拦洪区；

②滞洪水库或滞洪区。

（2）水库水位上升时，出流量随之增加。当洪水消退、进水量等于或小于出水量时，水位就不再上升。此后，水便自动地由水库泄出，直到放完多余蓄水为止。

2. Translate the following sentences into Chinese.

（1）A flood control reservoir or generally called flood－mitigation reservoir，stores a portion of the flood flows in such a way as to minimize the flood peaks at the areas to be protected downstream.

（2）A storage reservoir with gated spillway and gated sluiceway，provides more flexibility of operation，and thus gives us better control and increased usefulness of the reservoir.

Learning section

Expressions of Numbers 1（基数词的表示法）

基本基数词共31个，其中100以下的27个，100以上的4个。如表2.1所示。

Tab. 2. 1

100以下的基数词			100以上的基数词
1～10	11～19	20～90	
1 one	11 eleven		
2 two	12 twelve	20 twenty	
3 three	13 thirteen	30 thirty	100 a/one hundred
4 four	14 fourteen	40 forty	1000 a/ one thousand
5 five	15 fifteen	50 fifty	1000000 a/ one million
6 six	16 sixteen	60 sixty	1000000000 a/ one billion（美）
7 seven	17 seventeen	70 seventy	a/one thousand million（英）
8 eight	18 eighteen	80 eighty	
9 nine	19 nineteen	90 ninety	
10 ten			

（1）13～19数字的表达。

13～19的数字皆以－teen[－tiːn]结尾，其中，fourteen，sixteen，seventeen，eighteen和nineteen分别由four，six，seven，eight，nine加后缀－teen变成的，eighteen中只保留一个t。thirteen，fifteen分别由three和five转化而来。

（2）21～99数字的表达先说“几十”，再说“几”，中间要加连字号。

39 thirty－nine

（3）101～999数字的表达先说“几百”，再加and，再加末尾两位数或末位数

375 three hundred and seventy－five

（4）1000以上的数字的表达。

先从后向前数，每三位数前留一空，用这个方法把数目分作若干段，再一段段地念；自右向左第一个空前的数为thousand，第二个空前的数为million，第三个空前的数为billion，例：

8,021 eight thousand and twenty - one

13,849 thirteen thousand，eight hundred and forty - nine

631,562 six hundred and thirty - one thousand，five hundred and sixty - two

54,256,000 fifty - four million，two hundred and fifty - six thousand

970,000,000 nine hundred and seventy million

14,800,000,000 fourteen billion，eight hundred million

2,000,000,000,000 two trillion

Unit 2. Dam

Passage

A dam is a barrier that **impounds**[1] water or underground streams. Dams generally serve the primary purpose of retaining water, while other structures such as **floodgates**[2] or **levees**[3] (also known as **dikes**[4]) are used to manage or prevent water flow into specific land regions. Hydropower and pumped - storage hydroelectricity are often used in conjunction with dams to generate electricity. A dam can also be used to collect water or for storage of water which can be evenly distributed between locations.

The word *dam* can be traced back to Middle English, and before that, from Middle Dutch, as seen in the names of many old cities. Early dam building took place in Mesopotamia and the Middle East. Dams were used to control the water level, for Mesopotamia's weather affected the Tigris and Euphrates rivers, and could be quite unpredictable.

The earliest known dam is theJawa Dam in Jordan, 100 kilometres northeast of the capital Amman. This gravity dam featured a 4.5 m high and 1 m wide stone wall, supported by a 50 m wide earth **rampart**[5]. The structure is dated to 3000 BC. The Ancient Egyptian Sadd - el - Kafara Dam at Wadi Al - Garawi, located about 25 km south of Cairo, was 102 m long at its base and 87 m wide. The structure was built around 2800 or 2600 B.C. as a **diversion**[6] dam for flood control, but was destroyed by heavy rain during construction or shortly afterwards. By the mid - late third century BC, an **intricate**[7] water - management system within Dholavira in modern day India, was built. The system included 16 reservoirs, dams and various channels for collecting water and storing it.

Dams are classified on the basis of the type and materials of construction, as gravity, **arch**[8], **buttress**[9], and earth. The first three types are usually constructed of con-

[1]impound[im'paʊnd] *vt.* 将……关起来；扣押，监禁；搁置，保留；蓄水

[2]floodgate['flʌdgeit] *n.* （江河或湖泊的）防洪闸（门），（泄）水闸门

[3]levee['levi] *n.* 堤；早朝；[美]总统或其他高级官员所举行的招待会；专为某人举行的招待会

[4]dike[dɑik] *n.* 堤；排水沟；障碍物

[5]rampart['ræmpɑ:t] *n.* （城堡等周围宽阔的）防御土墙；防御，保护

[6]diversion[dai'vɜ:ʃn] *n.* 转移；分散注意力；消遣

[7]intricate['intrikət] *a.* 错综复杂的；难理解的；曲折；盘错

[8]arch[ɑ:tʃ] *n.* 弓形；拱门；拱形物；足弓，齿弓 *vt.* （使）弯成拱形；用拱连接，向后弯 *vi.* 拱起；成为弓形 *a.* 主要的；首要的；调皮的；淘气的

[9]buttress['bʌtrəs] *n.* 扶壁；支撑物 *vt.* 支持，鼓励；用扶壁支撑，加固

crete. A gravity dam depends on its own weight for stability and is usually straight in plan although sometimes slightly curved. Arch dams transmit most of the horizontal thrust of the water behind them to the **abutments**[10] by arch action and have thinner cross section than comparable gravity dams. Arch dams can be used only in narrow **canyons**[11] where the walls are capable of withstanding the thrust produced by the arch action. The simplest of the many types of the buttress dams is the slab type, which consists of **sloping**[12] flat slabs supported at intervals by buttress. Earth dams are **embankments**[13] of rock or earth with **provision**[14] for controlling seepage by means of an impermeable core or upstream blanket.

More than one type of dam may be included in a single structure. Curved dams may combine both gravity and arch action to achieve stability. Long dams often have a concrete river section containing **spillway**[15] and **sluice**[16] gates and earth or rock-fill wing dams for the remainder of their length.

The selection of the best type of dam for a given site is a problem in both engineering feasibility and cost. Feasibility is governed by topography, geology and climate. For example, because concrete **spalls**[17] when subjected to **alternate**[18] freezing and **thawing**[19], arch and buttress dams with thin concrete sections are sometimes avoided in areas subject to extreme cold. The relative cost of the various types of dams depends mainly on the availability of construction materials near the site and the accessibility of transportation facilities. Dams are sometimes built in stages with the second or later stages constructed a decade or longer after the first stage.

(672 words)

[10]abutment[əˈbʌtmənt] *n.* 邻接，桥墩，桥基；扶垛；对接；接界

[11]canyon[ˈkænjən] *n.* 峡谷

[12]sloping[ˈsləupiŋ] *a.* 倾斜的，有坡度的 *v.* 有斜度（slope 的现在分词）；悄悄地走；潜行

[13]embankment[imˈbæŋkmənt] *n.* 路堤；筑堤

[14]provision[prəˈviʒn] *n.* 规定，条项，条款；预备，准备，设备；供应，（一批）供应品；生活物质，储备物资 *vt.* & *vi.* 为……提供所需物品（尤指食物）

[15]spillway[ˈspilwei] *n.* 溢洪道，泄洪道；溢口

[16]sluice[slu:s] *n.* 水闸；（用水闸控制的）水；有闸人工水道；漂洗处 *vt.* 冲洗；（指水）喷涌而出；漂净；给……安装水闸

[17]spall[spɔ:l] *n.* （尤指岩石的）碎片，裂片 *vt.* & *vi.* 弄碎，击碎（矿石）

[18]alternate[ˈɔ:ltə:nət] *a.* 轮流的；交替的；间隔的；代替的 *vi.* 交替；轮流 *vt.* 使交替；使轮流 *n.* 〈美〉（委员）代理人；候补者；替换物

[19]thawing[ˈθɔ:iŋ] *n.* 熔化，融化 *v.* （气候）解冻（thaw 的现在分词）；（态度、感情等）缓和；（冰、雪及冷冻食物）溶化；软化

Exercises

1. Translate the following sentences into English.

（1）大坝可以被用来收集和储存水资源，以调节地区之间的水分配。

（2）拱坝通过拱效应把水的大部分水平推力传递到拱坝后两侧坝基上，他们相对重力坝具有更薄的截面积。

(3) 在具体地区选择最好类型的大坝不仅关系到工程可行性问题，还关系到经济问题。

2. Translate the following sentences into Chinese.

(1) Dams generally serve the primary purpose of retaining water, while other structures such as floodgates orlevees (also known as dikes) are used to manage or prevent water flow into specific land regions.

(2) Dams are classified on the basis of the type and materials of construction, as gravity, arch, buttress, and earth.

(3) A gravity dam depends on its own weight for stability and is usually straight in plan although sometimes slightly curved.

Learning section

Expressions of Numbers 2（分数的表示法）

分数词（Fractional Numerals）由基数词和序数词构成，基数词代表分子，序数词代表分母。除了分子为1的情况下，序数都要用复数的形式：

one - fourth　　five - ninths　　two - thirds　　three and two - fifths

seven - twelfths　　forty - seven and three - eighths

此外还有下面表示法：

a (one) half　　a (one) quarter　　three - quarters　　one and a half　　seven and three quarters

比较复杂的分数读法如下：

twenty - three over nine　　seventy - six over ninety - two

应注意的是：表达分数有时基数词与序数词之间需加“-”。“三分之二的学生”可写作 two - thirds of the students。

Unit 3. Gravity Dam

Passage

In a gravity dam, the force that holds the dam in place against the push from the water is Earth's gravity pulling down on the mass of the dam. The water presses laterally (downstream) on the dam, tending to overturn the dam by rotating about its toe (a point at the bottom downstream side of the dam). The dam's weight **counteracts**[1] that force, tending to rotate the dam the other way about its toe. The designer ensures that the dam is heavy enough that gravity wins that contest. In engineering terms, that is true whenever the resultant of the forces of gravity and water pressure on the dam acts in a line that passes upstream of the toe of the dam.

Furthermore, the designer tries to shape the dam so if one were to consider the part of dam above any particular height to be a whole dam itself, that dam also would be held in place by gravity. i. e. There is no tension in the upstream face of the dam holding the top of the dam down. The designer does this because it is usually more practical to make a dam of material essentially just **piled**[2] up than to make the material stick together against vertical tension.

Note that the shape that prevents tension in the upstream face also **eliminates**[3] a balancing compression stress in the downstream face, providing additional economy.

The designer also ensures that the toe of the dam is sunk deep enough in the earth that it does not slide forward.

For this type of dam, it is essential to have an **impervious**[4] foundation with high bearing strength.

When situated on a suitable site, a gravity dam can prove to be a better **alternative**[5] to other types of dams. When built on a carefully studied foundation, the gravity dam probably **represents**[6] the best developed example of dam building. Since the fear of flood is a strong **motivator**[7]

[1]counteract[ˌkaʊntərˈækt] *vt.* 抵消；阻碍；中和

[2]pile[pail] *v.* 堆起；堆叠；放置；装入

[3]eliminate[iˈlimineit] *vt.* 排除，消除；淘汰；除掉；[口]干掉

[4]impervious[imˈpɜːviəs] *a.* 不可渗透的；透不过的；无动于衷的；不受影响的

[5]alternative[ɔːlˈtɜːnətiv] *a.* 非正统的，不寻常的；两者择一的 *n.* 二中择一；可供选择的事物；取舍；非传统（或他择性）生活方式的追随者（或鼓吹者）

[6]represent[ˌrepriˈzent] *vt.* 表现，象征；代表，代理；扮演；作为示范 *vi.* 代表；提出异议

[7]motivator[ˈməʊtiveitə(r)] *n.* 激起行为（或行动）的人（或事物），促进因素，激发因素

in many regions, gravity dams are being built in some instances where an arch dam would have been more economical.

Gravity dams are classified as "solid" or "hollow" and are generally made of either concrete or **masonry**[8]. This is called "zoning". The core of the dam is zoned depending on the availability of locally available materials, foundation conditions and the material **attributes**[9]. The solid form is the more widely used of the two, though the hollow dam is frequently more economical to construct. Gravity dams can also be classified as "overflow" (spillway) and "non-overflow." Grand Coulee Dam is a solid gravity dam and Itaipu Dam is a hollow gravity dam.

(431 words)

[8]masonry['meisənri] *n.* 石工工程，砖瓦工工程；砖石建筑

[9]attributes[ə'tribju:ts] *n.* 属性，特性，特质；属性（attribute的名词复数）；（人或物的）特征；价值；[语法学]定语
v. 认为……是（attribute的第三人称单数）；把……归于；把……品质归于某人；认为某事（物）属于某人（物）

Exercises

1. Translate the following sentences into English.

（1）在细致研究的基础上，重力坝可能代表了大坝建设的先例。

（2）作用在坝上的水压力，通过绕坝址旋转作用试图把大坝推翻。

（3）设计者必须保证坝址在地下埋藏足够的深度以防止大坝向前倾倒。

2. Translate the following sentences into Chinese.

(1) In a gravity dam, the force that holds the dam in place against the push from the water is Earth's gravity pulling down on the mass of the dam.

(2) Furthermore, the designer tries to shape the dam so if one were to consider the part of dam above any particular height to be a whole dam itself, that dam also would be held in place by gravity. i. e.

(3) Note that the shape that prevents tension in the upstream face also eliminates a balancing compression stress in the downstream face, providing additional economy.

Learning section

Tab. 2. 2　　Expressions of Numbers 3

基本序数词的构成			
1st first	11th eleventh	**20th twentieth**	**30th thirtieth**
2nd second	**12th twelfth**	21st twenty - first	**40th fortieth**
3rd third	13th thirteenth	22nd twenty - second	**50th fiftieth**
4th fouth	14th fourteenth	23rd twenty - third	**60th sixtieth**
5th fifth	15th fifteenth	24th twenty - fourth	**70th seventieth**
6^{th}sixth	16th sixteenth	25th twenty - fifth	**80th eightieth**
7th seventh	17th seventeenth	26th twenty - sixth	**90th ninetieth**
8th eighth	18th eighteenth	27th twenty - seventh	100th one hundredth
9th ninth	19th nineteenth	28th twenty - eighth	$1,000^{th}$ one thousandth
10th tenth		29th twenty - ninth	$1,000,000^{th}$ one millionth
			$1,000,000,000^{th}$one billionth

（1）除“第一”、“第二”、“第三”之外，其他序数都以在基数词后加词尾 th 构成，其中有些词（见上表中的黑体部分）在拼法上有少许变化，例：fifth，eighth，ninth，twelfth，twentieth 等

（2）两位数的词，只需把个位数变为序数词：

第二十二 twenty - second，第七十八 seventy - eighth

（3）三位以上的词，只把最后的两位数变为序数词：

第九百九十九 nine hundred and ninety - ninth

此外，first，second 等词常可缩写，如 1st，2nd，3rd，20th，21^{st}等。

Unit 4. Embankment Dam

Passage

Embankment dams are made from **compacted**[1] earth, and have two main types, rock - fill and earth - fill dams. Embankment dams rely on their weight to hold back the force of water, like the gravity dams made from concrete.

Rock - fill dams are embankments of compacted free - draining **granular**[2] earth with an impervious zone. The earth utilized often contains a large percentage of large particles hence the term rock - fill. The impervious zone may be on the upstream face and made of **masonry**[3], concrete, plastic **membrane**[4], steel sheet piles, **timber**[5] or other material. The impervious zone may also be within the embankment in which case it is referred to as a core. In the instances where clay is utilized as the impervious material the dam is referred to as a **composite**[6] dam. To prevent internal erosion of clay into the rock fill due to seepage forces, the core is separated using a filter. Filters are specifically graded soil designed to prevent the **migration**[7] of fine grain soil particles. When suitable material is at hand, transportation is minimized leading to cost savings during construction. Rock - fill dams are resistant to damage from earthquakes. However, inadequate quality control during construction can lead to poor compaction and sand in the embankment which can lead to **liquefaction**[8] of the rock - fill during an earthquake. Liquefaction potential can be reduced by keeping **susceptible**[9] material from being saturated, and by providing adequate compaction during construction. An example of a rock - fill dam is New Melones Dam in California.

Earth - fill dams, also called earthen dams, rolled - earth dams or simply earth dams, are constructed as a simple embankment of well compacted earth. A **homogeneous**[10] rolled - earth dam is entirely constructed of one type of material but may contain a drain layer to collect seep water. A zoned - earth dam has distinct parts or zones of **dissimilar**[11]

[1] compact['kɒmpækt] v. 压紧，(使) 坚实

[2] granular['grænjələ(r)] a. 颗粒状的

[3] masonry['meisənri] n. 石工工程，砖瓦工工程；砖石建筑

[4] membrane['membrein] n. (动物或植物体内的) 薄膜；隔膜；(可起防水、防风等作用的) 膜状物

[5] timber['timbə(r)] n. 木材，木料；(用于建筑或制作物品的) 树木；用材林，林场；素质 vt. 用木料支撑；备以木材

[6] composite['kɒmpəzit] a. 混合成的，综合成的；[建]综合式的；[数]可分解的；[植]菊科的 n. 合成物，混合物，复合材料；[植]菊科植物

[7] migration[mai'greiʃn] n. 迁移，移居

[8] liquefaction[ˌlikwi'fækʃən] n. 液化

[9] susceptible[sə'septəbl] a. 易受影响的；易受感染的；善感的；可以接受或允许的

[10] homogeneous[ˌhɒmə'dʒiːniəs] a. 同性质的，同类的；由相同(或同类型) 事物 (或人) 组成的；均匀的；[数]齐性的，齐次的

[11] dissimilar[di'similə(r)] a. 不同的，不相似的

material, typically a locally plentiful shell with a **watertight**[12] clay core. Modern zoned - earth embankments employ filter and drain zones to collect and remove seep water and preserve the integrity of the downstream shell zone. An outdated method of zoned earth dam construction utilized a hydraulic fill to produce a watertight core. Rolled - earth dams may also employ a watertight facing or core in the manner of a rock - fill dam. An interesting type of temporary earth dam occasionally used in high latitudes is the frozen - core dam, in which a **coolant**[13] is circulated through pipes inside the dam to maintain a watertight region of **permafrost**[14] within it.

[12]watertight['wɔ:tətait] *a.* 不漏水的；水密的；防渗的；无懈可击的

[13]coolant['ku:lənt] *n.* 冷冻剂，冷却液，散热剂

[14]permafrost['pə:məfrɔst] *n.* （如极地的）永久冻土；多年冻土

Tarbela Dam is a large dam on the Indus River in Pakistan. It is located about 50 km northwest of Islamabad, and a height of 148 m above the river bed and a reservoir size of 250km^2 makes it the largest earth filled dam in the world. The principal element of the project is an embankment 2,700 metres long with a maximum height of 142 metres. The total volume of earth and rock used for the project is approximately 152.8 million cu. Meters, which makes it the largest man made structure in the world, except for the Great Chinese Wall which consumed somewhat more materials.

Because earthen dams can be constructed from materials found on - site or nearby, they can be very cost - effective in regions where the cost of producing or bringing in concrete would be **prohibitive**[15].

[15]prohibitive[prə'hibətiv] *a.* 禁止的；禁止性的；抑制的；（指价格等）过高的

(541 words)

Exercises

1. Translate the following sentences into English.

(1) 土石坝和混凝土重力坝一样，依靠他们的重量阻挡水的冲力。

(2) 均质碾压式土坝全部由一种类型的材质建成，但是可能包含一个排水层收集渗漏水。

(3) 由于建设土坝可以在场地或附近地区取材，因此在生产或引进混凝土价格过高的地区建设土坝具有更高的经济性。

2. Translate the following sentences into Chinese.

(1) Embankment dams are made from compacted earth,

and have two main types，rock－fill and earth－fill dams.

(2) Rock－fill dams are embankments of compacted free－draining granular earth with an impervious zone.

(3) Earth－fill dams，also called earthen dams，rolled－earth dams or simply earth dams，are constructed as a simple embankment of well compacted earth.

Learning section

Expressions of Numbers 4（小数的表示法）

6.4—six point four

0.8—zero（naught）point eight

0.05—（naught）point naught five 或 zero point zero five

0.726—（naught）point seven two six 或 zero point seven two six

12.409—twelve point four o nine

709.06—seven hundred and nine point o six

Unit 5. Spillway

Passage

A spillway is a section of a dam designed to pass water from the upstream side of a dam to the downstream side. Many spillways have floodgates designed to control the flow through the spillway.

A spillway is the safety valve for a dam. It must be designed to discharge maximum flow while keeping the reservoir below a **predetermined**[1] level. A safe spillway is extremely important. Many failures of dams have resulted from improperly designed spillways or spillways of insufficient capacity. Spillway size and frequency of use depend on the runoff characteristics of the drainage basin and the nature of the project. The determination and selection of the reservoir inflow design flood must be based on an adequate study of the hydrologic factors of the basin. The routing of the flow past the dam requires a reasonably **conservative**[2] design to avoid loss of life and property damage.

Site conditions greatly influence the location, type, and components of the spillway. The type of dam construction is also influenced by the type of spillway and spillway requirements.

There are six general categories of spillways: ① overflow, ②**trough**[3] or **chute**[4], ③side channel, ④**shaft**[5] or glory hole, ⑤ **siphon**[6], ⑥ gated. The designer may use one or a combination of types to fulfill the project needs.

Some designs will use one type of spillway for normal operation and for flood peaks up to a 50 - year or 100 - year frequency storm. An emergency spillway provides additional safety if emergencies arise that was not covered by normal design assumptions, such situations could result from floods above a certain level, malfunctioning spillway gates, or enforced shutdown of outlet works. The emergency spillway prevents **overtopping**[7] the main portion of the dam and is

[1]predetermine[ˌpriːdiˈtɜːmin] *vt.* & *vi.* 预先裁定；注定

[2]conservative[kənˈsɜːvətiv] *n.* 保守的人；（英国）保守党党员，保守党支持者 *a.* 保守的；（英国）保守党的；（式样等）不时新的

[3]trough[trɒf] *n.* 水槽，食槽；低谷期；[航]深海漕；[气]低气压槽

[4]chute[ʃuːt] *n.* 斜槽，滑道；降落伞 *vt.* 用斜槽或斜道运送 *vi.* 顺斜道而下，在斜槽或滑道中滑行

[5]shaft[ʃɑːft] *n.* 柄，轴；矛，箭；〈非〉嘲笑；光线 *vt.* 给……装上杆柄；〈俚〉苛刻的对待

[6]siphon[ˈsaifn]*n.* 虹吸管 *vt.* 用虹吸管吸（或输送）（液体）*vi.* 通过虹吸管

[7]overtop[ˈəʊvəˈtɒp] *vt.* 高出，高耸……之上

particularly needed for earth and rock embankments.

The overflow spillway is well suited to concrete dams. It is commonly used where dams have sufficient **crest**[8] length for the desired discharge capacity and where the foundation material is solid or can be protected against **scouring**[9]. Some dams use a free overflow or non - supported type; others incorporate a chute or trough to carry the flow to the downstream channel.

[8] crest [krest] *n.* 山顶；羽毛饰；鸡冠；(动物的) 颈脊 *vt.* 加上顶饰；到达……的顶部；形成冠毛状顶部；到达顶部
vi. 到达绝顶；形成浪峰；达到顶点

[9] scouring ['skaʊəriŋ] *n.* 擦(洗)净，冲刷，洗涤 *v.* 走遍 (某地) 搜寻 (人或物) (scour 的现在分词)；(用力) 刷；擦净；擦亮

Chute spillways are often used for earth dams or where there are poor downstream foundation materials. Side channels and shaft spillways are readily adapted to narrow canyons where space is limited. Limitations on crest length or maintaining a constant headwater level fit the flow characteristics of a siphon spillway. Gated spillways are used when it is desirable to limit the effects of the dam during high flows and prevent excessive flooding.

The spillway may be part of the dam or a separate structure. Its function must be integrated with the dam. The location, size, and other dam features influence the spillway location and arrangement. The final plan is governed by the overall economy and hydraulic sufficiency of the spillway.

The spillway can be gradually eroded by water flow, including **cavitation**[10] or turbulence of the water flowing over the spillway, leading to its failure. It was the inadequate design of the spillway which led to the 1889 over - topping of the South Fork Dam in Johnstown, Pennsylvania, resulting in the **infamous**[11] Johnstown Flood (the "great flood of 1889").

Erosion rates are often monitored, and the risk is ordinarily minimized, by shaping the downstream face of the spillway into a curve that minimizes turbulent flow, such as an **ogee**[12] curve.

(562 words)

[10] cavitation [ˌkævi'teiʃən] *n.* 空洞形成，气穴现象；空化；涡凹；汽蚀

[11] infamous ['infəməs] *a.* 声名狼藉的；无耻的，伤风败俗的；[法] (因犯重罪) 被剥夺公权的；很差的，低劣的

[12] ogee ['əʊdʒiː] *n.* 弯曲，S形；表反曲线

Exercises

1. Translate the following sentences into English.

(1) 溢洪道设计时必须能够保证泄放最大的流量，同时还能保持水库的水位在预定水位以下。

(2) 入库设计洪水的选择和确定，必须在充分研究流域水文因素的基础上进行。

(3) 通过汽蚀或溢洪道中水流的强烈冲击作用，都可能使溢洪道被侵蚀，由此可能引起溢洪道失去作用。

2. Translate the following sentences into Chinese.

(1) A spillway is a section of a dam designed to pass water from the upstream side of a dam to the downstream side.

(2) Spillway size and frequency of use depend on the runoff characteristics of the drainage basin and the nature of the project.

(3) An emergency spillway provides additional safety if emergencies arise that was not covered by normal design assumptions, such situations could result from floods above a certain level, malfunctioning spillway gates, or enforced shutdown of outlet works.

Learning section

to be in hot water/in deep water：陷入困境

keep one's head above water：不负债，免遭灭顶之灾

water over the dam：木已成舟，既成事实

Unit 6. Hydroelectric Power

Passage

Hydro means "water". So, hydropower is "water power" and hydroelectric power is electricity generated using water power. Potential energy (or the "stored" energy in a reservoir) becomes **kinetic**[1] (or moving energy). This is changed to mechanical energy in a power plant, which is then turned into electrical energy. Hydroelectric power is a renewable resource.

In an impoundment facility (see below), water is stored behind a dam in a reservoir. In the dam is a water **intake**[2]. This is a narrow opening to a tunnel called a **penstock**[3].

Water pressure (from the weight of the water and gravity) forces the water through the penstock and onto the **blades**[4] of a **turbine**[5]. A turbine is similar to the blades of a child's **pinwheel**[6]. But instead of breath making the pinwheel turn, the moving water pushes the blades and turns the turbine.

The turbine spins because of the force of the water. The turbine is connected to an electrical generator inside the powerhouse. The generator produces electricity that travels over long - distance power lines to homes and businesses. The entire process is called hydroelectricity.

1. Types of hydropower plants

There are three types of hydropower facilities: impoundment, diversion, and pumped storage. Some hydropower plants use dams and some do not. Many dams were built for other purposes and hydropower was added later. In the United States, there are about 80,000 dams of which only 2,400 produce power. The other dams are for **recreation**[7], stock/farm ponds, flood control, water supply, and irrigation.

Hydropower plants range in size from small systems for a home or village to large projects producing electricity

[1]kinetic[ki'netik] *a.* 运动的，活跃的，能动的，有力的；[物]动力（学）的，运动的

[2]intake['inteik] *n.* 吸入，进气；（液体等）进入口；摄入，摄取；纳入（数）量

[3]penstock['penstɒk] *n.*〈美〉水道；水渠；压力水管；水阀门

[4]blade[bleid] *n.* 刀片，剑；（壳、草等的）叶片；桨叶；浮华少年

[5]turbine['tɜːbain] *n.* 涡轮机；汽轮机；透平机

[6]pinwheel['pinwiːl] *n.* 轮转焰火，纸风车；针轮；靶中心

[7]recreation[ˌrekri'eiʃn]*n.* 消遣（方式）；娱乐（方式）；重建，重现

for utilities.

The most common type of hydroelectric power plant is an impoundment facility. An impoundment facility, typically a large hydropower system, uses a dam to store river water in a reservoir. Water released from the reservoir flows through a turbine, spinning it, which in turn activates a generator to produce electricity. The water may be **released**[8] either to meet changing electricity needs or to maintain a constant reservoir level.

A diversion, sometimes called run - of - river, facility channels a portion of a river through a canal or **penstock**[9]. It may not require the use of a dam.

When the demand for electricity is low, a pumped storage facility stores energy by pumping water from a lower reservoir to an upper reservoir. During periods of high electrical demand, the water is released back to the lower reservoir to generate electricity.

[8]release[ri'li:s] *vt.* 释放；放开；发布；发行
n. 释放，排放，解除；释放令；公映的新影片，发布的新闻[消息]

[9]penstock['penstɒk]
n.〈美〉水道；水渠；压力水管；水阀门

2. Sizes of hydroelectric power plants

Facilities range in size from large power plants that supply many consumers with electricity to small and micro plants that individuals operate for their own energy needs or to sell power to utilities.

Although **definitions**[10] vary, the U. S. Department of Energy defines large hydropower as facilities that have a capacity of more than 30 **megawatts**[11].

Although definitions vary, DOE defines small hydropower as facilities that have a capacity of 100 kilowatts to 30 megawatts.

A microhydropower plant has a capacity of up to 100 **kilowatts**[12]. A small or microhydroelectric power system can produce enough electricity for a home, farm, ranch, or village.

(511 words)

[10] definition [ˌdefi 'niʃn] *n.* 定义；规定，明确；[物]清晰度；解释

[11] megawatt ['megəwɒt] *n.* 兆瓦，百万瓦特（电能计量单位）

[12]kilowatt['kiləwɒt]
n.[电]千瓦

Exercises

1. Translate the following sentences into English.

（1）水压（水的自身重量和万有引力）驱使水渠中的水作用于水轮机上的叶片。

（2）就像吹气使纸风车转动一样，水流运动推动水轮机叶片并使其转动。

（3）水库中释放出来的水流经水轮机，使其旋转，最终转化为电能。

2. Translate the following sentences into Chinese.

（1）Hydropower is "water power" and hydroelectric power is electricity generated using water power.

（2）There are three types of hydropower facilities：impoundment，diversion，and pumped storage.

（3）Facilities range in size from large power plants that supply many consumers with electricity to small and micro plants that individuals operate for their own energy needs or to sell power to utilities.

Learning section

Tab. 2.3　Expressions of Measurement Units 1（长度单位）

简　写	汉语名称	英文名称	换　算
km	千米	kilometer	1km＝1000m
m	米	meter	1m＝1m
dm	分米	decimeter	1dm＝0.1m
cm	厘米	centimeter	1cm＝0.01m
mm	毫米	millimeter	1mm＝0.001m
mi	英里	mile	1mi＝1609.344m
ft	英尺	foot	1ft＝0.3048m
in	英寸	inch	1in＝0.0254m
yd	码	yard	1yd＝0.9144m

Unit 7. Curriculum for Hydraulic and Hydropower Engineering

Passage

Tab. 2. 4 Names and Credits for Compulsive Courses

NO.	Courses	Credits
1	Advanced Mathematics Ⅰ	10
2	Linear Algebra Ⅲ	2
3	College Physics	4
4	Engineering Drafting	3. 5
5	Engineering Survey	3
6	Theoretical **Mechanics**[1]	3. 5
7	Material Mechanics	3. 5
8	Construction Materials	2. 5
9	Hydraulics Ⅰ	3. 5
10	Hydraulic Engineering Geology	2. 5
11	Structural Mechanics	3. 5
12	Soil Mechanics	3. 5
13	Engineering Hydrology & Planning of Hydraulic Engineering & Hydraulic Energy	3. 5
14	**Reinforced Concrete Structure**[2] Theory	3. 5
15	Hydraulic Structures	4
16	Construction of Hydraulic Engineering	3
17	Structures of Hydropower Station	3
18	Experiments in College Physics	2
19	Experiments in Hydraulics	1
20	Practicum	1
21	Practice of Engineering Survey	2
22	Practice of Engineering Geology	1
23	Course Exercise of Engineering Hydrology and Planning of Hydraulic Engineering and Hydraulic Energy	1
24	Advanced Practicum	2
25	Course Exercise of Reinforced Concrete Structure Theory	1

[1] mechanics [mi 'kæniks] *n.* 力学

[2] Reinforced Concrete Structure 钢筋混凝土结构

续表

NO.	Courses	Credits
26	Course Exercise of Hydraulic Structures	1
27	Course Exercise of Hydropower Station Buildings	1
28	Course Exercise of Construction of Hydraulic Engineering	1
29	Undergraduate Practicum and Graduation Design/Dissertation	10

Tab. 2. 5 Names and Credits for Optional Courses

NO.	Courses	Credits
1	Probability Theory and Mathematical Statistics	2
2	Course of C Program Design	3
3	Project **Budget**[3] and **Tender**[4]	2
4	Introduction of Profession	1. 5
5	Principles and Applications of Database	2
6	Electro－technics and Electrical Equipments	2
7	AutoCAD	2
8	Hydraulics Ⅱ	2
9	Hydraulic Calculation in **Cold Regions**[5]	2
10	Rock Mechanics	2
11	Professional English	2
12	Structural Mechanics Ⅱ	2
13	Theory of Steel Structure	2
14	Engineering Project Management	2
15	Economy of Water Conservancy Project	2
16	**Frost Resistance**[6] Techniques of Hydraulic Structure	2
17	Workshop Structures of Hydropower Station	2
18	Flood Control Scheduling of Reservoir	2
19	Theory of Project Supervision	2
20	Hydraulic Engineering Management	2
21	Hydraulic Structure Safety Testing	2
22	Computational Method	2
23	**Geosynthetics**[7]	2
24	Elasticity and Finite Element Method	2. 5
25	Bridge Engineering	2
26	Road Engineering	2
27	Outline of Water Administration	2
28	**Entrepreneur**[8] Management	2
29	Hydraulic Machinery	2
30	Techniques of Roller Compacted Concrete	2

[3]budget['bʌdʒit] *n.* 预算；预算额，经费 *vt.* & *vi.* 编制预算，安排开支等

[4]tender['tendə] *n.* 投标 *a.* 脆弱的，幼弱的；温柔的；亲切的；疼痛的，敏感的

[5]cold region 寒区

[6]frost resistance 抗冻性

[7]geosynthetic[ˌdʒiə'sin'θetik] *n.* 土工合成材料

[8]entrepreneur[ˌɔntrəprə'nəː] *n.* 企业家，创业，创业者

Exercises

1. Build sentences with following words.

(1) curriculum，hydraulic and hydropower engineering

(2) hydraulics，mechanics，structure

(3) techniques，professional English，computer skills

2. Translate the following sentences into English.

(1) 水利水电工程的核心课程有：理论力学、材料力学、结构力学、土力学、水利工程测量、工程水文学、水力学、水利工程项目管理、水工建筑物、水电站建筑物、水利水能规划、水利工程施工等。

(2) 水利水电工程专业课程由必修课和选修课组成，其中概率论与数理统计、C语言程序设计、AutoCAD、专业英语等都是比较重要的选修课。

Learning section

Tab. 2.6 Expressions of Measurement Units 2（面积单位）

简 写	汉语名称	英文名称	换 算
km^2	平方千米	square kilometer	$1km^2=10^6m^2$
m^2	平方米	square meter	$1m^2=1m^2$
dm^2	平方分米	square decimeter	$1dm^2=10^{-2}m^2$
cm^2	平方厘米	square centimeter	$1cm^2=10^{-4}m^2$
mm^2	平方毫米	square millimeter	$1mm^2=10^{-6}m^2$
sq. mi·	平方英里	square mile	—
sq. ft	平方英尺	square foot	—
sq. in	平方英寸	square inch	—
sq. yd	平方码	square yard	—
ha	公顷	Hectare	$1ha=10^4m^2$

Unit 8. Disciplines Distribution of Hydraulic and Hydropower Engineering in China

Passage

Tab. 2. 7　　Doctoral degree programs of Hydraulic and Hydropower Engineering (15 in total)

1	Tsinghua University
2	Dalian University of Technology
3	Hohai University
4	Wuhan University
5	Sichuan University
6	Xi'an University of Technology
7	Tianjin University
8	China Institute of Water Resources and Hydropower Research
9	Zhengzhou University
10	Nanjing Hydraulic Research Institute
11	Northwest Agriculture and Forest University
12	Huazhong University of Science and Technology
13	Xinjiang Agricultural University
14	Ningxia University
15	Yangzhou University

Tab. 2. 8　　Master degree programs of Hydraulic and Hydropower Engineering (42 in total)

Province	NO.	Affiliation
Anhui	1	HeFei University of Technology
Beijing	4	Beijing University of Technology
		Tsinghua University
		China Agricultural University
		China Institute of Water Resources and Hydropower Research
Gansu	2	Gansu University of Technology
		Lanzhou Jiaotong University
Hebei	1	Agricultural University of Hebei
Henan	2	North China University of Water Resources and Electric Power
		Zhengzhou University

续表

Province	NO.	Affiliation
Heilongjiang	1	Heilongjiang University
Hubei	5	Changjiang River Scientific Research Institute
		Huazhong University of Science and Technology
		China Three Gorges University
		Wuhan University
		China University of Geosciences
Hunan	1	Changsha University of Science and Technology
Jilin	1	Jilin University
Jiangsu	4	Hohai University
		Jiangsu University
		Nanjing Hydraulic Research Institute
		Yangzhou University
Jiangxi	1	Nanchang University
Liaoning	2	Dalian University of Technology
		Shenyang Agricultural University
Inner Mongolia	1	**Inner Mongolia**[1] Agricultural University
Ningxia	1	Ningxia University
Shandong	3	Shandong University
		Shandong Agricultural University
		Ocean University of China
Shanxi	1	Taiyuan University of Technology
Shaanxi	2	Xi'an University of Technology
		Northwest Agriculture and Forest University
Shanghai	1	Tongji University
Sichuan	1	Sichuan University
Tianjin	1	Tianjin University
Tibet	1	**Tibet**[2] University
Xinjiang	1	Xinjiang Agricultural University
Yunnan	1	Kunming University of Science and Technology
Zhejiang	1	Zhejiang University

[1] Inner Mongolia 内蒙古

[2] Tibet[ti'bet] *n.* 西藏

Exercises

1. Build sentences with following words.

(1) hydraulic and hydropower engineering，university，institute

(2) the Three Gorges，the Yangtze River

(3) Inner Mongolia Agricultural University，Tibet University，Tsinghua University

2. Translate the following sentences into English.

(1) 水利水电工程是我国重要的基础设施和基础产业。

(2) 水利水电工程专业以水利枢纽（水坝、水闸、水电站等）为主要对象，学生主要学习水利水电工程建设所必需的数学、力学和工程结构、水利水能规划、水利工程经济计算等方面的基本理论和基本知识。

(3) 水利水电工程学生需要掌握必要的工程设计方法、施工管理方法和科学研究方法，具有水利水电工程及相关工程勘测、规划、设计、施工、科研和管理等方面的基本能力。

Learning section

Tab. 2. 9　Expressions of Measurement Units 3（体积单位）

简　写	汉语名称	英文名称	换　算
m^3	立方米	cubic meter	$1m^3=1000L$
cm^3	立方厘米	cubic centimeter	$1cm^3=1mL$
L	升	litre	1L=1000mL
mL	毫升	mililitre	$1mL=1cm^3$
gal.	加仑	gallon	1gal. = 3.785L

Part 3

Figure 1. Basic Photosynthesis

Exercises

1. Build sentences with following words.

(1) photosynthesis, plants

(2) the sun, evaporation

(3) oxygen, carbon dioxide

2. Translate the following sentences into English.

(1) 光合作用是植物、藻类和某些细菌，在可见光的照射下，将二氧化碳和水转化为有机物，并释放出氧气的过程。

(2) 阳光、水、二氧化碳是光合作用的三要素。

3. Try to describe the process of basic photosynthesis in English.

Learning section

to be in hot water/in deep water：陷入困境

keep one's head above water：不负债，免遭灭顶之灾

[1]photosynthesis[ˌfəutəu'sinθisis] *n.* 光合作用，光能合成

[2] oxygen ['ɔksidʒən] *n.* 氧，氧气

[3]carbon['kɑːbən]*n.* 碳

[4] dioxide [dai'ɔksaid] *n.* 二氧化物

[5]carbon dioxide 二氧化碳

Figure 2. Irrigation System on the Hill

[1] conventional irrigation 传统灌溉

[2] drip irrigation 滴灌

[3] dynamic[daiˈnæmik] *a.* 有活力的，强有力的；不断变化的；动力的，动态的

[4] drawdown[ˈdrɔːdaun] *n.* （抽水后）水位降低，水位降低量

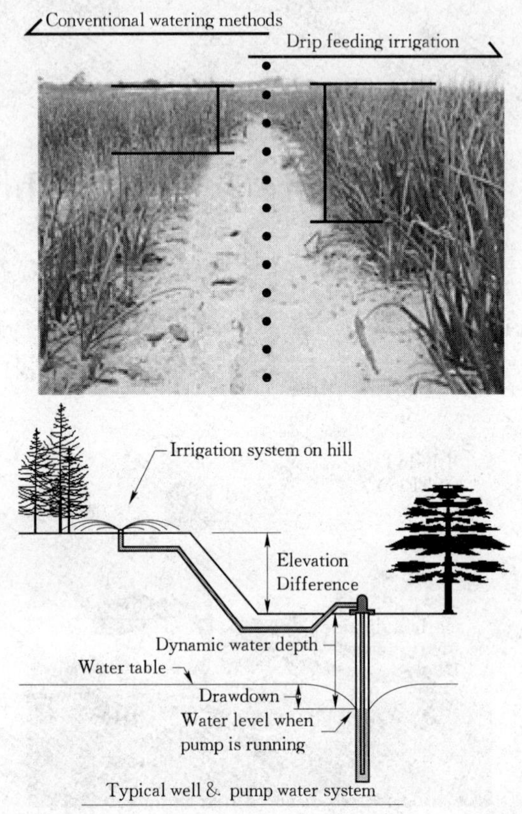

Exercises

1. Build sentences with following words.

(1) conventional irrigation, drip irrigation

(2) elevation, water level

(3) dynamic water depth, drawdown

2. Translate the following sentences into English.

(1) 在平原区滴灌比传统灌溉效率更高；而在山区山坡灌溉比传统灌溉效率更高。

(2) 在水井抽水时，地下水位会发生变化。

3. Try to describe the process of irrigation system on the hill in English.

Learning section

water over the dam：木已成舟，既成事实

to hold water：（理论、计划等）证明合理，说得通

Unit 1. Water and Agriculture

Passage

China's **drought**[1] disasters occur frequently. Drought in the spring is very serious to the north of the Qinling Mountain Divide and Huaihe River with a frequency of nine in ten years, sometimes these droughts are even **accompanied**[2] with summer droughts or summer to autumn droughts. Summer droughts or **consecutive**[3] summer to autumn droughts are very common in the middle and lower reaches of the Yangze River, dominated with summer droughts; consecutive spring to summer droughts are very common in the upper reach of the Yangze River and the southwestern region, dominated with spring droughts, or with summer droughts or autumn droughts in some regions.

Drought disasters have caused huge losses to agricultural production. The losses of grain production due to drought disasters account for more than 50% of the total grain production losses caused by various natural disasters, and account for approximately 5% of the total amount of average annual grain production.

China can be divided into three different irrigation zones based on **precipitation**[4]: **perennial**[5] irrigation zone where the average annual precipitation is less than 400mm; unstable irrigation zone where the average annual precipitation is greater than 400mm but less than 1000mm; and **supplemental**[6] irrigation zone where the average annual precipitation is greater than 1000mm.

The conditions for agricultural production are not very satisfactory in most parts of China, especially the threats from drought and without irrigation in regions where the annual precipitation is less than400mm; irrigation plays a vital role in guaranteeing good harvests in **semi - humid**[7] and semi - arid regions; seasonal drought turn out to be greater harmful to agriculture in southeastern coastal areas, as a re-

[1] drought[draʊt]*n.* 旱灾；干旱（时期），旱季

[2]accompany[ə'kʌmpəni] *n.* 伴随……同时发生；陪伴；陪同

[3]consecutive[kən'sekjʊtiv] *a.* 连续的，连贯的

[4]precipitation [pri,sipi'teiʃ(ə)n]*n.* 降雨量

[5]perennial[pə'reniəl]*a.* 终年的，长久的；多年生的；

[6]supplemental[,sʌpli'mentəl]*a.* 补充的，追加的

[7]semi - humid 半潮湿，半湿性

sult irrigation is also a prerequisite to guarantee agricultural production.

China's irrigation **schemes**[8] can be divided into three big **categories**[9], i. e. water storage, water lift and water division. According to the investigation, the per unit area yield of grain is approximately 6000kg per hectare in areas without irrigation facilities. The yield of grain from irrigated farm is 2 – 4 times of that from non – irrigated one. Moreover, there will be more yield increase in areas that is drier. According to the analysis, 70% of the national total grain production, 80% of the national total cotton production and more than 90% of the national total cotton production and more than 90% of the national total vegetable production come from **arable**[10] land with irrigation **facilities**[11]. **Apparently**[12], irrigation is a key and major measure to **enhance**[13] the capacities to cope with drought disasters as well as to **guarantee**[14] good harvest.

(406 words)

[8] schemes [skiːmz] *n.* 计划（scheme 的名词复数）；体系；阴谋

[9] category [ˈkætig(ə)ri] *n.* 种类，类别；派别

[10] arable [ˈærəb(ə)l] *a.* 适用于耕种的 *n.* 耕地

[11] facility [fəˈsiliti] *n.* 设备；设施；能力

[12] apparently [əˈpærəntli] *ad.* 显然地；表面上；显而易见

[13] enhance [inˈhæns] *vt.* 提高；增加；加强

[14] guarantee [gær(ə)nˈtiː] *vt.* 保证；担保

Exercises

1. Build sentences with following words.

(1) drought disaster

(2) irrigation facilities

(3) irrigation scheme

2. Translate the following sentences into English.

(1) 根据降水情况，中国分为三个不同的灌溉地带。

(2) 中国大部分地区农业生产条件不十分理想，特别是干旱缺水对农业的威胁最大。

3. Translate the following sentences into Chinese.

(1) Drought disasters have caused huge losses to agricultural production. The losses of grain production due to drought disasters account for more than 50% of the total grain production losses caused by various natural disasters, and account for approximately 5% of the total amount of average annual grain production.

(2) According to the investigation, the per unit area yield of grain is approximately 6000kg per hectare in areas without irrigation facilities. The yield of grain from irrigated

farm is 2 - 4 times of that from non - irrigated one.

Learning section

Tab. 3. 1 Expressions of Measurement Units 4（温度单位）

简　写	汉语名称	英文名称	换　算
℃	摄氏度	Centigrade Degree	摄氏度＝（华氏度 － 32）÷1.8
℉	华氏度	Fahrenheit	华氏度＝32＋摄氏度×1.8

Unit 2. Irrigation Methods

Passage

Irrigation methods can be divided into four main types - surface irrigation, subsurface irrigation, sprinkler irrigation, and drip irrigation and many subtypes. Surface irrigation is the oldest type and still accounts for about three fourths of all irrigation. Subsurface irrigation is limited in its **adaptation**[1]. Sprinkler irrigation can be used in any climate, is the most popular method in humid regions, and is still expanding in use. Drip irrigation makes the most efficient use of water.

[1]adaptation[ædəp'teiʃ(ə)n] *n.* 适应，顺应；同化

Surface Irrigation

Surface irrigation is the oldest and most used method of irrigation. Farmers in China, Egypt, India, and countries of the Middle East are known to have irrigated lands at least 4000 years ago, most likely using surface methods. Surface irrigation includes **border irrigation**[2], **furrow irrigation**[3], **basin irrigation**[4] and **water spreading**[5].

1. Border Irrigation

Border irrigation makes used of **parallel**[6] earth **ridges**[7], called **borders**[8], to guide a sheet of flowing water across a field. The area between two borders is the border **strip**[9]. These strip may vary from 3 to 30 meters (10 to 100 feet) in width and from 100 to 800 meters (300 to 2600 feet) in length.

[2]border irrigation 畦灌
[3]furrow irrigation 沟灌
[4] basin irrigation 淹灌
[5] water spreading 漫灌
[6] parallel['pærəlel] *n.* 平行线(面)；相似物 *a.* 平行的
[7] ridge[ridʒ] *n.* 田埂；背脊，峰；隆起线；山脊
[8] border['bɔːdə] *n.* 畦；边；镶边；边界
[9] strip[strip] *n.* 长条；带状地带

Border irrigation can be used with slope gradients between 0.2% and 2% for cultivated crops, up to 4% or 5% for small grain or hay crops, and up to about 8% for **pastures**[10]. Extensive land leveling is often required because the **topography**[11] must be smoother than for furrow irrigation. The cost of land leveling is offset by the low labor requirement for turning water into a few border rather than into many furrows or **corrugations**[12]. The smooth topography is easy to work across at harvest time.

[10] pasture['pɑːstʃə]*n.* 牧草地，牧场
[11]topography[tə'pɒgrəfi]*n.* 地形；地貌；地形学
[12] corrugation[ˌkɔːrʊ'geiʃən] *n.* 沟；车辙

2. Furrow Irrigation

Furrow irrigation is used with row crops by running water in the cultivated between the **rows**[13]. The rows can be fed with **siphon**[14] tubes channel or in groups from ditch turnouts. Furrows are particular suitable for irrigation crops that are subject to **injury**[15] if water covers the crown or stems of the plants. Row crops such as vegetables, cotton, corn, **maize**[16], potatoes, and seeds crops planted on raised beds are irrigated by furrows placed between the plant rows. **Orchards**[17] vineyard can be irrigated by placing one or more furrows between the tree or vine rows and in order to wet a major area of the root zone.

3. Basin Irrigation

Basin irrigation is probably the oldest method. It was practiced in Egypt more than 5000 years ago. It is a simple method that is still widely used to keep land flooded for long periods for paddy rice production or for shorter periods for many other crops.

In basin irrigation, the field to be irrigated is divided into units surrounded by small levees or dikes. A ditch or other water supply large enough to flood the basin must be available on one side. Water is turned in until the desired depth is reached, then cut back to just enough to hold constant depth of about 10cm for paddy rice or shut off completely for other crops. The water in the basin may be allowed to completely infiltrate or, in some low **permeability**[18] soil, the excess may be drained onto a lower basin after a specified time.

4. Water Spreading

Water spreading involves turning a stream of water onto a relatively flat field and allowing the water to spread naturally. This is normally a very inefficient method of irrigation providing little or no control of water distribution over the field. Water spreading is sometimes employed in low-lying areas with a stream or small river where water can be diverted during periods of high water, such as associated with spring runoffs, and used for pre-irrigation.

Subsurface Irrigation

Subsurface irrigation, also called sub-irrigation, can

[13] row[rəʊ]*n.* 行；排；路，街；吵闹

[14] siphon['saif(ə)n]*n.* 虹吸管

[15] injury['in(d)ʒ(ə)ri]*n.* 伤害；损害；伤害的行为

[16] maize[meiz]*n.* 玉米；黄色，玉米色

[17] orchard['ɔːtʃəd]*n.* 果园；果园里的全部果树

[18] permeability[pɜːmiə'biliti]*n.* 渗透性；磁导率；可渗透性

[19]ditch[ditʃ]n. 沟渠；壕沟

[20] coarse[kɔːs]a. 粗糙的；粗鄙的

be considered as a controlled drainage system. **Ditches**[19] are usually used, but some systems use tile lines. The systems remove water during wet seasons and add it during dry seasons so the water table is always at a controlled depth. That depth might be as little as 30cm for shallow - rooted vegetation in a **coarse**[20] sandy soil or as great as 120cm in some loamy soils. The surface soil should be dry but most of the root zone should be moist. The field can even be cultivated and irrigation at the same time.

(704 words)

Exercises

1. Build sentences with following words.

(1) border irrigation

(2) furrow irrigation; subsurface irrigation

(3) basin irrigation; water irrigation

2. Translate the following sentences into English.

(1) 采用畦灌的地形比采用沟灌的地形要平坦，因此，常常需要在大面积范围内进行土地平整。

(2) 地下灌溉系统通常使用明沟，有些工程也使用暗管。

3. Translate the following sentences into Chinese.

(1) The systems remove water during wet seasons and add it during dry seasons so the water table is always at a controlled depth. That depth might be as little as 30 cm for shallow - rooted vegetation in a coarse sandy soil or as great as 120 cm in some loamy soils.

(2) Basin irrigation is probably the oldest method. It was practiced in Egypt more than 5000 years ago.

Learning section

Tab. 3. 2 Expressions of Measurement Units 5 (时间单位)

简 写	汉语名称	英文名称	换 算
a	年	year	1a=12mon=365d
mon	月	month	1mon=28－31d
d	日	day	1d=24h
h	时	hour	1h=60min
min	分	minute	1min=60sec
sec	秒	second	——

Unit 3. Drip Irrigation and Drip Irrigation System

Passage

Drip Irrigation

Drip irrigation is the one that achieves the highest irrigation efficient: about 90% of the applied water is available to the plant. High efficiency is achieved by supplying water to individual plants through small plastic lines. Water is supplied either continuously or so frequently that the plant roots grow in constantly moist soil.

Drip irrigation is especially suitable for watering trees or other large plants. Much of its use has been in **orchards**[1] and **vineyards**[2] but it has also been used to irrigate a variety of row crops including several kinds of vegetables and fruits. Its advantages are greatest where areas between plants can be left dry. It has no advantage for close-growing vegetation such as lawns, pastures, or small grain crops.

An Israeli engineer named Symcha Blass developed the idea of drip irrigation in the 1930s, but it had to wait until plastic tubing was available to make a practical system. Drip irrigation in the United States increased from 40**ha**[3] in 1960 to over 50,000ha in 1976. Nearly half of the drip irrigation in the United States is in California, some of it in **avocado**[4] orchards with slopes up to 50% or 60%. Erosion is not a problem because there is no runoff.

A **bonus**[5] with drip irrigation is its ability to use water with a higher salt content than any other method up to about 2500mg/**liter**[6]. The constant flow of water from the drip **emitter**[7] toward the outer **edges**[8] of the plant root zone carriers the salt along with it. Salt concentrations become very high in the dry areas between plants but not in the actual root zone.

Drip irrigation saves water, is able to use water high salt, functions well in all but the **extremes**[9] of coarse and

[1] orchard[ˈɔːtʃəd]*n.* 果园；果园里的全部果树

[2] vineyard[ˈvinjɑːd]*n.* 葡萄园

[3] ha[hɑː]*n.* 面积公顷（hectare）缩写，我国不推荐使用，在我国常使用 hm^2

[4] avocado[ˌævəˈkɑdo]*n.* 鳄梨，鳄梨树；暗黄绿色

[5] bonus[ˈbəʊnəs]*n.* 奖金，额外津贴；红利；额外令人高兴的事情

[6] liter[ˈlitər]*n.* 公升（容量单位）

[7] emitter[iˈmitə]*n.* 发射器；这里指滴灌灌水器滴头

[8] edge[edʒ]*n.* 边缘；锋利，尖锐

[9] extreme[ikˈstriːm]*a.* 极端的，过激的；极限的，非常的

fine textured soils, works on almost any **topography**[10] without causing **erosion**[11], and required little labor. The disadvantages are mainly high equipment costs and plugging of the lines by **sediment**[12], salt **encrustation**[13], or large.

Drip Irrigation System

A drip irrigation system normally includes a **control box**[14] that **regulates**[15] the water pressure, **filters**[16] the water, and provides for the addition of **fertilizers**[17] and **herbicides**[18]. Chorine may be added to **eliminate**[19] **algal**[20] growth. The water pressure for drip irrigation is normally 0.4 to $1kg/cm^2$ as compared to 1 to $8kg/cm^2$ for sprinkler irrigation. Some drip controls are set to increase the pressure periodically and flush the lines to reduce clogging.

Drip irrigation lines **branch**[21] into several parts at three or four stages to provide the many outlets required. The last stage is a flexible plastic **lateral**[22] line 12 to 32mm in diameter that lies either on or just below the soil surface and applies the water either through small holes in the line or through emitter **nozzles**[23]. Emitter nozzles lead the water through a long spiral path that slows the flow and permits a larger emission hole to be used. The larger hole is less subject to **plugging**[24].

Several **components**[25] of typical drip systems are shown in Fig. 3.1. Water is pumped into most systems and flow through valves, filters, mainlines, sub-mains or manifold lines, and laterals before it discharged into the field through point-source emitters, **bubblers**[26], or microsprinkers. Sometimes line sources of water are obtained with either **porous**[27] tubes that discharge water continuously along their length or tubes that

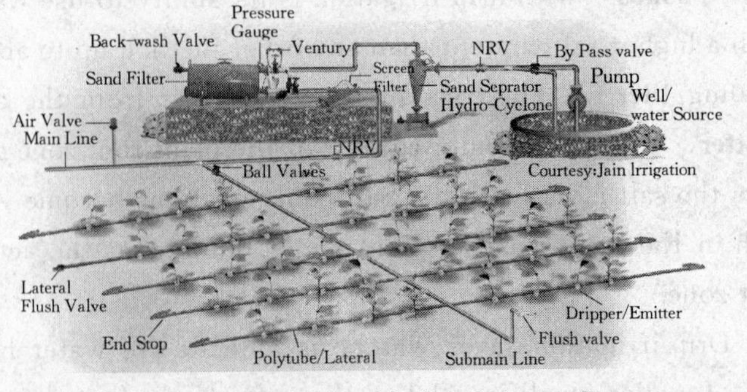

Fig. 3.1 The Components of a Drip Irrigation System

[10] topography[tə'pɒgrəfi] n. 地形

[11] erosion [i'rəʊʒ(ə)n] n. 侵蚀，腐蚀，磨损

[12] sediment ['sedim(ə)nt] n. 沉积物，沉渣

[13] encrustation[ɛn'krʌs'teʃən] n. 硬壳；结壳；用覆盖物

[14] control box 首部枢纽

[15] regulate['regjʊleit] vt. 调节，调整；控制，管理

[16] filter['filtə] vt. 过滤；透过；渗透

[17] fertilizer['fɜːtilaizə] n. 肥料，化肥

[18] herbicide['hɜːbisaid] n. 除草剂

[19] eliminate[i'limineit] vt. 排除，消除；淘汰，除掉

[20] algal['ælgəl] a. 海藻的

[21] branch [brɑːn(t)ʃ] n. 树枝，分枝；分部；支流；vt. 分支；出现分歧；vi. 分支；出现分歧

[22] lateral ['læt(ə)r(ə)l] a. 侧面的，横向的，这里 lateral line 指毛管

[23] nozzle ['nɒz(ə)l] n. 管嘴，喷嘴

[24] plug[plʌg] vi. 填塞，堵

[25] component[kəm'pəunənt] n. 成分；组分；零件

[26] bubblers ['bʌb(ə)lə(r)] n. 喷水式饮水口

[27] porous['pɔːrəs] a. 多孔渗水的；能渗透的；有气孔的

discharge water through closely spaced openings instead of laterals with point - source e-mitters or microsprinklers. Filiters are sometimes omitted in bubbler and spray sprinkler systems. Drip system may or may not include chemical injection equipment.

(581 words)

Exercises

1. Build sentences with the following words.

(1) drip irrigation system

(2) the advantages of drip irrigation

2. Translate the following sentences into English.

(1) 与其他灌水方法相比，滴灌还有一个优点，即它可以使用含盐量更高的水源。

(2) 滴灌的缺点主要是设备费用高，管道易被泥沙、盐分结晶或藻类植物堵塞。

3. Try to describe the components of a drip irrigation system in English.

Learning section

Tab. 3. 3 Expressions of Measurement Units 6（质量单位）

简　写	汉语名称	英文名称	换　算
t	吨	ton	1t=1000kg
kg	千克	kilogram	1kg=1000g
g	克	gram	1g=10mg
mg	毫克	milligram	—
lb	磅	pound	1lb=0. 454kg
oz	盎司	ounce	1lb=16oz
ct	克拉	carat	1ct=0. 2g

Unit 4. Water-saving Irrigation Development

Passage

1. Water Saving Irrigation in Ancient and Recent Times

Ancient China had already applied the water saving technology into agricultural production, such as land leveling, narrowing furrow and basin size for irrigation, and deep **plowing**[1] to loosen the soil. In the early 1930s, Pangshan Irrigation Experimental Site was established in Wujiang County of Jiangsu Province and Fenghuai District Experimental Site was established in Linhuaigang of Anhui Province to conduct scientific experiment on saving irrigation water use on farm lands.

2. Water Saving Irrigation from1950s to 1980s

Chinese water saving irrigation in this period mainly focused on research and **extension**[2] of individual technology. Planned water use, canal **seepage**[3] control and improved furrow and border irrigation technology et al centering on increasing irrigation water use efficiency were conducted; hundreds of irrigation experimental stations were established and the irrigation scheduling of some major crops was brought forward; "seeping paddy field with new methods and irrigation **paddy**[4] field with shallow water" was extended in the south; improved furrow and basin irrigation technology was advanced in the north. Land leveling and constructing horizontal **terraced**[5] field were popularized throughout the country to realize the overall planning and **comprehensive**[6] management of hills, water, farmlands, forests and roads.

In the 1970s, internationally advanced sprinkler irrigation and micro irrigation technologies were imported and digested into China while gradually extending canal seep control, and some cheap mechanical and automatic irrigation equipment were developed in China independently. Low pressure pipeline irrigation was stressed and extended in the 1980s.

3. Water Saving Irrigation since the 1990s

Chinese water saving irrigation has entered into a rapid development stage since the1990s. The government at various levels had

[1]plowing[plauiŋ] n.[农学]翻耕，耕作 v. 耕地（plow 的 ing 形式）；犁

[2]extension[ik'stenʃ(ə)n] n. 延长；延期；扩大；伸展；电话分机

[3] seepage ['si:pidʒ] n. [流] 渗流；渗漏；渗液

[4] paddy['pædi] n. 稻田（复数 paddies）；爱尔兰人；Patrick（男子名）和 Patricia（女子名）的昵称

[5] terraced['terəst] a. 阶地的；有平台的；沿斜坡建造的；v. 使成阶地；使成梯田（terrace 的过去式和过去分词）

[6]comprehensive [kɒmpri'hensiv] n. 综合学校；专业综合测验 a. 综合的；广泛的；有理解力的

strengthened the organization and guidance for water saving irrigation and offered specific **discounted**[7] **loans**[8] and financial assistance funds to intensify technical guidance and services. In addition to this, a lot of water saving and thousands of water saving irrigation technology demonstration plots had been established, and the main works of all of the large irrigation districts and parts of the medium irrigation districts had been rehabilitated.

(332 words)

[7] discounted ['diskauntid]a. 已贴现的；已折扣的 v. 打折扣 (discount 的过去分词)；不重视

[8] loans[lons]n.［金融］贷款（loan 的复数形式）；借贷

Exercises

1. Translate the following sentences into English.

（1）在20世纪50～80年代，中国的节水灌溉多偏重单项技术的研究和推广。围绕提高灌溉用水效率开展了计划用水、渠道防渗、改进沟畦灌溉技术等工作。

（2）20世纪90年代以来，中国的节水灌溉进入了快速发展阶段。

2. Try to summarize water saving irrigation in ancient and recent times in English.

Learning section

Tab. 3. 4　Expressions of Measurement Units 7（力的单位）

简　写	汉语名称	英文名称	换　算
N	牛顿	Newton	—
kN	千牛	kilo newton	1kN=1000N

Unit 5. Water - saving Irrigation Measures

[1]temporal['temp(ə)r(ə)l] a. 时间的；世俗的；暂存的

[2] spatial ['speiʃ(ə)l]a. 空间的；存在于空间的；受空间条件限制的

[3]conjunctive[kən'dʒʌŋ(k)tiv] a. 连接的

[4] rationally['ræʃənli]ad. 讲道理地，理性地

[5] brackish['brækiʃ]a. 有盐味的，可厌的

[6] arid['ærid]a. 干旱的，干燥的；贫瘠的，荒芜的，不毛的

[7] urban ['ɜːb(ə)n]a. 都市的

[8] aggregate['ægrigət]vt. 使聚集，使积聚

[9] asphalt['æsfælt]n. 沥青，柏油；（铺路等用的）沥青混合料

To Rationally Allocate Irrigation Water Sources

Adjusting the **temporal**[1] and **spatial**[2] distribution of various water sources in river basins and regions through structural and non - structures and conductive water allocation among multiple water sources so that to achieve the maximum irrigation benefits.

In the surface water irrigation district, "reservoir groups strung together by canals" type irrigation system can be developed to realize **conjunctive**[3] water allocation among large, medium, small, water storage, water diversion and water lift schemes. In the well irrigation district, reducing groundwater exploitation and **rationally**[4] exploiting and using **brackish**[5] water to realize irrigation directly using brackish water or using mixed fresh water and brackish water. In **arid**[6] hilly regions, rainwater can be harvested by various means, such as slope surface and road, which can harvest rainfall efficiently, and then be diverted and stored into cellars and pools as crucial water for crops to fight against drought. Wastewater discharged from **urban**[7] domestic uses or industries should be treated to a standard that meet irrigation water quality. This kind of water can be used to irrigate non - immediate - edible crops.

To Increase Water Conveyance and Allocation Efficiency

Canal lining are mainly used in surface water irrigation districts. Earth material, **aggregated**[8] rock, coating materials and concrete or **asphalt**[9] concrete materials can be used to line canals to form impervious layer so that to increase irrigation water conveyance and allocation efficiency. Water delivery with low pressure pipeline system is mainly applied in well irrigation districts. Pipes such as embedded concrete pipes, ground nylon - coated or vinylon - coated hoses, or ground plastic membrane hoses can be used to deliver water to increase the irrigation water conveyance and allocation efficiency.

In order to meet requirements of irrigation system, to facilitate the general public's production, living and transportation as well as to facilitate water allocation and measurement, there is a need to **rehabilitate**[10] the gate, **culverts**[11], bridges, **aqueducts**[12] and **hydraulic drops**[13] et canal system structures so that to increase the irrigation water conveyance and allocation efficiency; there is also a need to test the efficiency and to rehabilitate the **devices**[14] of water - lifting pumping stations.

Agronomic Water Saving Technologies

Agronomic[15] water conservation is not only an indispensable component of the technical system of water saving irrigation, but also a main measure for efficient use in rained agriculture. The measures mainly include: to apply **drought - resistant**[16] variety; to apply soil moisture conservation technology by **tillage**[17] and land cover; to apply water - fertilizer **coupling**[18] technology, and to apply chemical agent. Out of which, the coupling model and technology focusing on fertilizer, water and crop yield can increase the fertilizer use efficiency by 3%-5% and increase the yield by 20%-30% under conditions of not increasing the amount of use of fertilizer and water.

Management Water Saving Technologies and Measure

The water saving technologies and measures in the aspect of management mainly include: total amount control and **quota**[19] management in irrigation water use, water saving irrigation scheduling, soil moisture monitoring and irrigation forecasting technology, water measurement technology in irrigation district, and automatic monitoring and control technology in irrigation district et al.

(505 words)

[10] rehabilitate[ri:hə'biliteit] *vt.* 使康复；使恢复名誉；使恢复原状 *vi.* 复兴；复权；恢复正常生活

[11] culverts['kʌlvəts] *n.* [交] 涵洞；暗沟（culvert 的复数）

[12] aqueduct['ækwidʌkt] *n.* [水利] 渡槽；导水管；沟渠

[13] hydraulic drops 水力降；[水利]跌水

[14] devices[di'vaisis] *n.* [机][计]设备；[机]装置；[电子]器件（device 的复数）

[15] agronomic[,ægrə'nɑmik] *a.* 农事的；农艺学的

[16] drought - resistant 抗旱的

[17] tillage['tilidʒ] *n.* 耕作，耕种

[18] coupling['kʌpliŋ] *n.* [电]耦合；结合，联结 *v.* 连接（couple 的 ing 形式）

[19] quota['kwəʊtə] *n.* 配额；定额；限额

Exercises

1. Build sentences with following words.

(1) low pressure pipeline; allocation

(2) canal lining

2. Translate the following sentences into English.

(1) 发展“长藤结瓜”式灌溉系统，实现“大、中、小、蓄、引、提”联合调度。

（2）农艺节水既是节水灌溉技术体系不可缺少的组成部分，也是雨养农业高效用水的主要措施。主要包括：选用抗旱品种、应用耕作蓄水保墒技术、覆盖保墒技术、水肥耦合技术和施用化学制剂等。

3. Try to describe the main measures of water saving irrigation in English

Learning section

throw cold water：泼冷水

A lot of/much water has flown/run under the bridge since：自从那时起，已经过了很久

Unit 6. Land Drainage

Passage

Land drainage removes excess surface water from an area or lowers the groundwater below the root zone to improve plant growth or reduce the accumulation of soil salt. Land - drainage systems have many **features**[1] in common with **municipal**[2] storm drainage of surface water, at a considerable saving in cost over that of buried pipe. Open ditches, which are less **objectionable**[3] in rural areas than in cities, are widely used under suitable soil conditions ditches may also serve to lower the water table. However, closely spaced open ditches will **interfere**[4] with farm operations, and the more common method of draining excess soil water is by use of buried drains. Drains usually empty into ditches, although the modern tendency is to use large pipe in **lieu**[5] of ditches where possible. This frees extra land for cultivation and does away with **unsightly**[6] and sometimes dangerous open ditches. Since land drainage is normally a problem in very flat or leveled land, a **disposal**[7] work provided with tide gates and pumping equipment is often necessary for the final removal of the collected water.

Land drainage is not as demanding in terms of **hydrologic**[8] design as other type of drainage. The purpose of land drainage is to remove a volume of water in a reasonable time. Where sub - drainage is installed to remove excess water from irrigated land for land salinity control, the volume of leaching water to be applied in each irrigation is known; and the drains should be capable of removing this volume in the interval twice irrigation.

A ditch - drainage system consists of **laterals**[9], submains, and main ditches. Ditches are usually **unlined**[10]. Small ditches may be constructed with special ditching machine, while larger ditches are often **excavated**[11] with a **dragline**[12]. Some very large

[1] feature[ˈfiːtʃə] n. 特征，特点；容貌

[2] municipal[mjʊˈnisip(ə)l] a. 市的，市政的；地方自治的；都市的

[3] objectionable [əbˈdʒekʃ(ə)nəb(ə)l] a. 令人不快的，讨厌的

[4] interfere[intəˈfiə] vi. 干预，干涉；调停，排解；妨碍，打扰

[5] lieu [ljuː; luː] n. 代替；场所，处所

[6] unsightly[ʌnˈsaitli] a. 不美观，难看的，不好看的

[7] disposal[diˈspəʊz(ə)l] a. 处理（或置放）废品的

[8] hydrologic n. 水文学

[9] laterals[ˈlæt(ə)r(ə)ls] n. 侧根；侧面部分（lateral 的复数）

[10] unlined[ˌʌnˈlaind] a. [服装] 无衬里的

[11] excavated [ˈɛkskəˌvetid] v. 发掘；挖掘（excavate 的过去式，过去分词）

[12] dragline[ˈdræglain] n. 牵引绳索；[机] 拉铲挖土机；绳斗电铲

ditches are contrasted with floating **dredge**[13]. Unless the excavated material (spoil) is needed as a levee to provide additional flow area in the ditch, it should be placed at least 15 feet (4.5m) back from the edge of the ditch so that its weight does not contribute to the **instability**[14] of the ditch bank. Spoil bank decreases the cultivated area and prevents inflow of water from the land adjacent to the ditch.

The slope available for drainage ditches is small; the cross section should approach the most efficient section as closely as possible. A **trapezoidal**[15] cross section is most common, with side slopes not **steeper**[16] than 1 on 1.5. Slopes of 1∶2 or 1∶3 are required in sandy soils. Occasionally, where the drainage water must be pumped, ditches are deliberately made with an inefficient section to create as much storage as possible to minimize peak pumping loads. The slope, **alignment**[17], and spacing of ditches are determined mainly by local topography.

Land drainage speeds up the **runoff**[18] of water and, hence, increase peak flow **downstream**[19] of the drained area. The consequences of this increase should be considered in the planning of drainage system. Wetlands are important biological areas. They serve migratory waterfowl and, in coastal areas, as nursery grounds for many important commercial species of aquatic life. The consequence of draining such lands requires careful evaluation.

(516 words)

[13] dredge[dredʒ]*n.* 挖泥船，疏浚机；拖捞网 *vt.* (用挖泥船等)疏浚；(用拖捞网等)捞取；(在食物上)撒(面粉等) *vi.* 疏浚，挖掘；采捞

[14] instability[instəˈbiliti]*n.* 不稳定(性)；基础薄弱；不安定

[15] trapezoidal[ˌtræpiˈzɔidəl]

a. [数] 梯形的；不规则四边形的

[16] steeper[ˈstiːpə]*n.* 浸润器；浸泡用的桶子

a. 陡峭的，险峻的

[17] alignment[əˈlainm(ə)nt]

n. 队列，成直线；校准；结盟

[18] runoff[ˈrʌnɒf]*n.* [水文] 径流；决赛；流走的东西

a. 决胜的

[19] downstream[ˈdənnˈstriːm]

a. 下游的；顺流的

ad. 下游地；顺流而下

Exercises

1. Translate the following sentences into English.

(1) 排水系统是由干沟、分干沟和支沟组成。排水系统通常是不衬砌的。小型沟道可以用专用钩机进行施工，而较大的沟道通常用拉铲挖土机施工。

(2) 农田排水的目的是为了在一定时间内排除一定体积的水。因此，为了控制盐分，设置地下排水系统来排除灌溉土地上多余的水分。

2. Translate the following sentences into Chinese.

(1) Land drainage remove excess surface water from an area or lowers the groundwater below the root zone to improve

plant growth or reduce the accumulation of soil salt.

(2) The slope available for drainage ditches is small; the cross section should approach the most efficient section as closely as possible. A trapezoidal cross section is most common, with side slopes not steeper than 1 on 1.5. Slopes of 1∶2 or 1∶3 are required in sandy soils.

Learning section

above water：脱离困境的

as unstable as water：像水一样不稳定，变化无常

Unit 7. Disciplines Distribution of Water Conservancy in China

Passage

Tab. 3. 5 Total numbers of undergraduate[1] majors which have been set belong to the first level discipline[2] of water conservancy[3]

Majors	1994. 4	2001. 3	2002. 12
Hydrology & Water Resources Engineering	6	20	29
Hydraulics and River Dynamics	23	36	41
Hydraulic Structure Engineering	9	14	16
Hydraulic & Hydropower Engineering	20	27	—
Harbor[4], Coastal and Offshore Engineering[5]	49	175	218

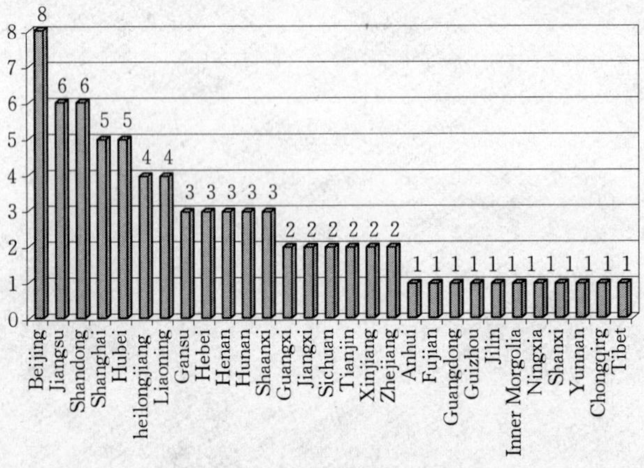

Fig. 3. 2 Distribution of **postgraduate**[6] education institutions of the disciplines of water conservancy

Tab. 3. 6 First level doctoral degree programs[7] of water conservancy (10 in total)

No. 7th	Tsinghua University	Beijing
	Dalian University of Technology	Dalian
	Hohai University	Nanjing
	Wuhan University	Wuhan
No. 8th	Sichuan University	Chengdu
	Xi'an University of Technology	Xi'an

[1] undergraduate[ˌʌndəˈgrædjuit] *n.* （未获学士学位的）大学生，大学肄业生

[2] discipline[ˈdisiplin] *vt.* 训练，训导；处罚，惩罚
n. 训练，锻炼，训导；纪律；处罚，处分，学科

[3] conservancy[kənˈsəːvənsi] *n.* （自然物源的）保护，管理，水土保持

[4] harbor[ˈhaːbə] *n.* 海港

[5] Coastal and Offshore Engineering 海岸与近海工程

[6] postgraduate[pəustˈgrædʒuːeit] *n.* 研究生 *a.* 研究生的

[7] doctoral degree program 博士学位授予点，博士学位学科点

续表

No. 9th	Tianjin University	Tianjin
	China Institute of Water Resources and Hydropower Research	Beijing
No. 10th	Zhengzhou University	Zhengzhou
	Nanjing Hydraulic Research Institute	Nanjing

[8] State Key Laboratory 国家重点实验室

Tab. 3. 7 State Key Laboratories[8] of Water Conservancy

State Key Laboratory of Water Resources & Hydropower Engineering Science	Wuhan University
State Key Laboratory of Coastal & Offshore Engineering	Dalian University of Technology
State Key Laboratory of Hydraulics & Mountain River Engineering	Sichuan University
State Key Laboratory of Hydrology & Water Resources and Hydraulic Engineering Science	Hohai University
State Key Laboratory of Hydroscience and Engineering	Tsinghua University
State Key Laboratory of Ocean Engineering	Shanghai Jiaotong University
State Key Laboratory of **Turbulence**[9] and **Complex**[10] System	Peking University
State Key Laboratory of **Estuarine**[11] & Coastal Research	East China Normal University
State Key Laboratory of Lake Science and Environment	Nanjing Institute of Geography and **Limnology**[12], Chinese Academy of Sciences
State Key Laboratory of Frozen Soil Engineering	Cold and Arid Regions Environmental and Engineering Research Institute, Chinese Academy of Sciences
Resources and State Key Laboratory of Environmental Information System	Institute of Geographic Sciences and Natural Resources Research

[9] turbulence['tə:bjuləns] *n.* 气体或水的涡流；波动

[10] complex['kɔmpleks] *a.* 由许多部分组成的，复合的；复杂的，难懂的

[11] estuarine['estjuərin] *a.* 河口的，江口的

[12] limnology [lim'nɔlədʒi] *n.* 湖泊学

ঔExercises ঔ

1. Build sentences with following words.

(1) institution，university，state key laboratory

(2) undergraduate，postgraduate，and doctoral students

(3) costal，offshore，ocean，river

2. Translate the following sentences into English.

(1) 随着高等教育事业以及水利事业的迅速发展，水利类本科专业和学科设置发展很快，为水利事业源源不断地输送人才。截止 2005 年，全国设有水利类专业的高等学校、水利高等院校和科研院所共 76 所。

(2) 对于毕业后打算继续深造的同学，了解我国水利工程学科的发展动态是重要且必要的。

(3) 水利工程一级学科包括 5 个专业：①水文与水资源工程，②水力学与河流动力学，③水工结构工程，④水利水电工程，⑤港口海岸和近海工程。

(4) 在中国有 76 家水利类硕士研究生培养单位，其中黑龙江有 4 家，陕西 3 家，内蒙古 2 家，西藏 1 家。

(5) 中国同时拥有水利类一级博士点和水利类国家级重点实验室的单位有 5 家：①清华大学，②大连理工大学，③河海大学，④武汉大学，⑤四川大学。

(6) 中国没有水利一级博士点，而拥有水利类国家级重点实验室的单位有 3 家：①北京大学湍流与复杂系统国家级重点实验室，②上海交通大学海洋工程实验室，③华东师大河口海岸学实验室。

ঔLearning section ঔ

as weak as water：身体非常虚弱；意志薄弱；性格懦弱

back water：退缩；死气沉沉的状态；文化落后的地方

Unit 8. Curriculum for Agricultural Water Conservancy Engineering

Passage

Tab. 3. 8 Names and Credits for Compulsive Courses

NO.	Courses	Credits
1	Advanced Mathematics Ⅰ	10
2	Linear Algebra Ⅲ	2
3	College Physics	4
4	Engineering Drafting	3. 5
5	Engineering Survey	3
6	Theoretical Mechanics	3. 5
7	Material Mechanics	3. 5
8	Construction Materials	2
9	Hydraulics Ⅰ	3. 5
10	Engineering Geology and Hydrogeology	2
11	Structural Mechanics	3. 5
12	Soil Mechanics	3. 5
13	Engineering Hydrology	2
14	Reinforced Concrete Structure Theory	3. 5
15	Hydraulic Structures	2. 5
16	Construction of Hydraulic Engineering	2
17	Irrigation and Drainage Engineering	2
18	**Pedology**[1] and Crop Science	2
19	Planning and Management of Water and Soil Resources	2
20	**Pump and Pumping Station**[2]	2
21	Experiments in College Physics	2
22	Practicum	1
23	Practice of Engineering Survey	2
24	Practice of Engineering Geology	1
25	Advanced Practicum	2
26	Course Exercise for Reinforced Concrete Structure Theory	1
27	Course Exercise of Hydraulic Structures	1
28	Course Exercise of Pump and Pumping Station	1
29	Course Exercise of Irrigation and Drainage Engineering	1
30	Graduation Design / Dissertation	10

[1] pedology [pi ˈdɔlədʒi] *n.* 土壤学

[2]Pump and Pumping Station 水泵与泵站

[3] drainage ['dreinidʒ] n. 排水；污水

Tab. 3. 9　Names and Credits for Optional Courses

No.	Courses	Credits
1	Probability Theory and Mathematical Statistics	2
2	Course of C Program Design	3
3	Project Budget and Tender	2
4	Introduction of Profession	1
5	Principles and Applications of Database	2
6	Electro - technics and Electrical Equipment	2
7	Irrigation and **Drainage**[3] Engineering Ⅱ	1
8	History of Water Conservancy in China	1
9	Economy of Water Conservancy Project	2
10	AutoCAD	2
11	Hydraulics Ⅱ	1
12	Hydraulic Calculation	2
13	Professional English	2
14	Theory of Steel Structure	2
15	Engineering Project Management	2
16	Groundwater Exploitation and Utilization	2
17	Agricultural Environment Science	2
18	Techniques and Applications of 3S	2
19	Computational Method	2
20	Outline of Water Administration	2
21	Water Supply and Drainage Engineering of Cities and Towns	2
22	Introduction of Bridge Engineering	2
23	Introduction of Soil and Water Conservation	2
24	Entrepreneur Management	2
25	Techniques of Water Saving Irrigation	2
26	Hydraulic Engineering Management	2
27	Small Hydropower Station	2

Exercises

1. Build sentences with following words.

(1) curriculum, hydraulic and hydropower engineering

(2) hydraulics, mechanics, structure

(3) techniques，professional English，computer skills

2. Translate the following sentences into English.

(1) 农业水利工程的核心课程有：灌溉排水工程学、土壤学及农作物学、水土资源规划与管理、水泵及水泵站、水利工程测量、工程地质与水文地质、水利工程施工、农村水电站、环境水利学等。

(2) 农水专业的课程可分为必修课和选修课，其中专业英语为选修课。

(3) 必修课分为基础课、课设、实习课三类，如钢筋混凝土、钢混课设、钢混实习。

(4) 农水专业毕业生应掌握农业水利工程勘测、规划、设计、施工、管理和科学研究等方面的专业技能，能从事农业和城市水利工程方面工作的高级工程技术人才。毕业生可在农业水土工程、水土保持与沙漠化防治、水文学及水资源、水力学与河流动力学等专业方向继续深造。

Learning section

bring the water to one's mouth (=make one's mouth water)：使某人垂涎

Don't cast out the foul water till you bring in the clean (=Don't throw out your dirty water until you get in fresh)：清水未来，莫泼脏水

Part 4

Unit 1. Snow Cover

[1] signify['signifai] vi. 表示；意味；预示
[2] transition[træn'ziʃ(ə)n] n. 过渡；转变；[分子生物] 转换；变调
[3] aided['eidid] v. 帮助（aid的过去分词）
[4] radiation[reidieiʃ(ə)n] n. 辐射；发光；放射物
[5] albedo[æl'bi:dəʊ] n.（行星等的）反射率；星体反照率
[6] absorption[əb'zɔ:pʃ(ə)n] n. 吸收；全神惯注，专心致志
[7] Alaskan 阿拉斯加州人的
[8] interior [in'tiəriə] n. 内部；本质
[9] tripod['traipɒd] n. [摄] 三脚架；三脚桌
[10] anchor['æŋkə] vt. 抛锚；使固定；主持节目
[11] the Tanana River 塔纳纳河
[12] Nenana 尼纳纳（城市名）
[13] onset['ɒnset] n. 开始，着手；发作；攻击，进攻
[14] consecutive[kən'sekjʊtiv] a. 连贯的；连续不断的
[15] the Southeast Panhandle 阿拉斯加东南的狭长地带
[16] climatology[klaimə'tɒlədʒi] n. 气候学；风土学

Passage

In much of the state, snow is on the ground at least half the year, from autumn to spring. Most areas experience a gradual increase in snow depth throughout the winter with a maximum sometime in spring. Snowmelt generally begins in mid to late spring and **signifies**[1] a relatively quick period of **transition**[2] that is **aided**[3] by an increase in daylight hours and the amount of incoming solar **radiation**[4]. As the snowmelt occurs and bare ground appears, there is a lowering of the surface **albedo**[5], or reflectivity. This results in more **absorption**[6] of solar radiation, leading to further temperature increases. **Alaskans**[7] refer to this time as breakup because the disappearance of snow is accompanied by the breakup of river and lake ice. In the **interior**[8], residents have watched for a particular local sign of spring since 1917. A **tripod**[9] is **anchored**[10] in the ice of **the Tanana River**[11] at **Nenana**[12]. The time at which the tripod falls in the river as the ice melts is considered the official breakup time.

There is regional variability to the **onset**[13] and melting of the seasonal snowpack, or time of continuous snow cover. Observations show that the number of **consecutive**[14] days with snow on the ground is highest in the Arctic and lowest in the southern coastal regions, especially **the Southeast Panhandle**[15]. This is, of course, related to the temperature and snowfall **climatologies**[16] of these regions.

The seasonal snowpack is established in early September for the Arctic coast and gradually increases until early spring.

Snowmelt normally occurs in late May or early June. By mid June the ground is normally snow - free. The average snow depth at **Barrow**[17] usually stays under 10 inches (25 cm) throughout the winter season and reaches a maximum depth in spring . **Recall**[18], however, that the average seasonal snowfall is 30 inches (76 cm). Snow depth is lower than snowfall because of a phenomenon called **metamorphosis**[19] of snow, in which the snow **crystal**[20] structure changes as **compaction**[21] occurs and the snowpack **density**[22] increases. Factors such as temperature and wind affect the rates of these processes.

In the Interior, the seasonal snowpack is established in early October, with the depth steadily increasing until a maximum snow depth of 20 to 30 inches (51 cm to 76 cm) is reached in February and March. Snowmelt normally takes place in April, with a complete disappearance in early May. Maximum snow depths can reach over 50 inches (127 cm) in spring, with minimum depths for the same period of less than 10 inches (25 cm) . This is similar to the seasonal snowpack features of **the Copper River Basin**[23]; however, the snow depth is normally not as high. The start and end of the snow season is slightly longer in the West Central region than the Interior and Copper River Basin and maximum spring snow depths are lower at 15 inches (38 cm) or less.

Along the Southcentral coast in areas that receive high seasonal snowfall totals, such as **Valdez**[24], the seasonal snow cover starts in late October or early November, with a disappearance by May. Average snow depths gradually increase, usually until March when the maximum of 45 to 50 inches (114 to 127 cm) is reached. This is quite different from other **maritime**[25] areas in southern Alaska such as **Adak**[26] and **Cold Bay**[27] in the **Aleutians**[28] or **Juneau**[29] and **Ketchikan**[30] in the Southeast Panhandle. Despite monthly10 to over 20 inches (25 to 51 cm), the mean monthly snow depth in these regions is less than 5 to 10 inches (13 to 25 cm) . Temperatures during the winter are such that enough melting takes place to keep the seasonal snowpack relatively low.

(626 words)

[17] Barrow 巴罗（美国）
[18] recall[riˈkɔːl] *n.* 召回；回忆；撤销
[19] metamorphosis [ˌmetəˈmɔːfəsis] *n.* 变形；变质
[20] crystal[ˈkristl] *a.* 水晶的；透明的，清澈的
[21] compaction[kəmˈpækʃən] *n.* 压紧；精简；密封；凝结
[22] density[ˈdensəti] *n.* 密度
[23] the Copper River Basin 铜河流域
[24] Valdez 瓦尔迪兹（美国）
[25] maritime[ˈmæritaim] *a.* 海的；海事的；沿海的；海员的
[26] Adak 埃达克（美国）
[27] Cold Bay 科尔德湾（美国）
[28] Aleutians 阿留申群岛
[29] Juneau 朱诺（美国）
[30] Ketchikan 凯奇坎（美国）

Exercises

1. Build sentences with the following words.

(1) Alaskans, breakup, river and lake ice

(2) factors, affect, processes

2. Translate the following sentences into English.

(1) 这当然是与这些地区的气温和降雪气候有关。

(2) 融雪现象通常从4月开始，5月初积雪彻底消失。

3. Translate the following sentences into Chinese.

(1) Snowmelt generally begins in mid to late spring and signifies a relatively quick period of transition that is aided by an increase in daylight hours and the amount of incoming solar radiation.

(2) Temperatures during the winter are such that enough melting takes place to keep the seasonal snowpack relatively low.

Learning section

环北极国家

the United States	美国
Canada	加拿大
Russia	俄罗斯
Denmark	丹麦
Finland	芬兰
Sweden	瑞典
Norway	挪威
Iceland	冰岛

Unit 2. Permafrost

passage

Definition of Permafrost

A widely accepted definition of permafrost is that it is ground that has a temperature low than 0℃ (32℉) continuously , for at least two **consecutive**[1] years. Much permafrost is thousands of years old, and some is of recent origin. Permafrost is forming now in places previously unfrozen: under thickening moss of boreal forests, and in the ground on the north side of new buildings (in the northern **hemisphere**[2]) where the midday sun no longer shines. However, the twentieth century has been largely an era of global warming so, by and large, the amount of permafrost is in decline. Each winter the ground in many places freezes to depths of 1 or 2 meters but no permafrost forms because the following summer's warmth brings the temperature up to 0℃ or above.

[1] consecutive[kən'sekjʊtiv] *a.* 连贯的；连续不断的

[2] hemisphere ['hemisfiə] *n.* 半球

This definition of permafrost on the basis of the ground's temperature alone has a **generality**[3] that may be useful in certain circumstances since it includes material such as rock that might contain no water at all. Nevertheless, my experience is that a person is hard pressed to avoid thinking of permafrost as anything other than soil that remains frozen year around; that is, to think of permafrost as a substance rather than a condition—and anyone who has tried to dig up permafrost knows that it is a tough substance indeed.

[3] generality[dʒenə'ræliti] *n.* 概论；普遍性；大部分

The word "permafrost" (permanent＋frozen) also is a catch - all term used to describe the general area of science relating to all aspects of frozen or nearly frozen ground, past, present and future. **Geocryology**[4] (earth ＋ cold ＋ study) also is a term for this field, and yet another in use is **periglacial**[5] (near ＋ glacier) processes.

[4] geocryology [dʒi:əukri'ɔlədʒi] *n.* 冻土地貌学

[5] periglacial[ˌperi'gleiʃ(ə)l] *n.* 冰缘，冰边

International Symposiums on Permafrost Engineering

The series of International **Symposiums**[6] on Permafrost Engineering are organized by the Melnikov Permafrost Institute

[6] symposium[sim'pəuziəm] *n.* 讨论会，座谈会；专题论文集；酒宴，宴会

(**Yakutsk**[7], Russia), the Cold and Arid Regions Engineering and Environmental Research Institute (Lanzhou, China), and the Heilongjiang Institute of Cold Region Engineering (Harbin, China) with support of the Russian Academy of Sciences Council on Earth Cryology and the International Permafrost Association. Since the first meeting in **Chita**[8] in 1993, the symposiums have been held **alternately**[9] in Russia and China (the 2nd in Harbin in 1996, the 3rd in Chita in 1998, the 4th in Lanzhou in 2000, the 5th in Yakutsk in 2002, the 6th in Lanzhou in 2004, the 7th in Chita in 2007, the 8th in Xi'an in 2009, and the 9th in **Mirny**[10] in 2011. The next, 10th, symposium on permafrost engineering will be held in Harbin, China in 2014.

The purpose of this symposium series is to provide a **forum**[11] for discussion of permafrost engineering issues, as well as for exchange of practical experience in construction and **maintenance**[12] of engineering structures on frozen ground. It aims to bring together researchers and engineers who work in cold/permafrost regions to discuss the ways and means for early detection of **adverse**[13] frost-related processes that may cause economic losses and environmental damage. Special attention is given to development of climate change adaptation measures.

The Symposium Proceedings are published by the organizers for each symposium.

(545 words)

[7] Yakutsk 雅库茨克

[8] Chita 赤塔

[9] alternately[ɔːlˈtɜːnətli] *ad.* 交替地；轮流地；隔一个地

[10] Mirny 米尔内（萨哈共和国）

[11] forum[ˈfɔːrəm] *n.* 论坛，讨论会；法庭；公开讨论的广场

[12] maintenance[ˈmeintənəns] *n.* 维护，维修；保持；生活费用

[13] adverse[ˈædvɜːs] *a.* 不利的；相反的；适对的

Exercises

1. Build sentences with the following words.

(1) the twentieth century, global warming, in decline

(2) special attention, development, climate change adaptation measures

2. Translate the following sentences into English.

(1) 许多冻土存在了数千年，有些则是最近才出现。

(2) 下一届，即第10届，冻土工程会议将于2014年在中国哈尔滨召开。

3. Translate the following sentences into Chinese.

(1) This definition of permafrost on the basis of the

ground's temperature alone has a generality that may be useful in certain circumstances since it includes material such as rock that might contain no water at all.

(2) The purpose of this symposium series is to provide a forum for discussion of permafrost engineering issues, as well as for exchange of practical experience in construction and maintenance of engineering structures on frozen ground.

Learning section

draw water in/with a sieve (=pour water into a sieve)：竹篮打水一场空

draw water to one's own mill：处处为自己打算；谋取私利

Unit 3. Freshwater and Permafrost in Alaska

Passage

Freshwater

Freshwater, in both liquid and frozen form , is abundant in Alaska , which has more than 3 million lakes, 12000 rivers, **innumerable**[1] streams , **creeks**[2] , and **ponds**[3] , as well as thousands of square miles of **glaciated**[4] area. **Lake Iliamna**[5] is the largest lake in the state at 1150 square miles (2978sq. km); it is located in southwest Alaska , east of **Bristol Bay**[6] . Two - thirds of the States is drained by the Yukon River and its **tributaries**[7]. This river enters from **the Yukon Territory**[8] at Alaska' eastern border and flows through the central part of the mainland. The longest river in Alaska , the Yukon empties into **the Bering Sea**[9] after a 1400 - mile (2253 - kilometer) journey through the state include the **Colville**[10] , Copper , **Kuskokwim**[11] , and **Noatak**[12] rivers.

Fun fact: These rivers and many other **navigable**[13] waterways are an important part of Alaska's history, as they provided an essential mode of transportation during the period of early exploration. Even today, much of Alaska is not **accessible**[14] by road , and river travel—whether by boat on open water in summer or by **sled**[15] over ice in winter—remains important for many villages. Rivers are a primary means for shipping supplies during the summer. In winter , rivers and streams provide an open area to **navigate**[16] by dog team when traveling between villages or out to winter trapping sites. The **salmon**[17] runs on many rivers also provide an important food source.

The glaciated area in Alaska covers about 29000 square miles (75110km^2), an area larger than some U. S. states (Molina 2001) . Several types of glaciers are found in Alaska: **tidewater**[18] glaciers, **piedmont**[19] glaciers , mountain glaciers , valley glaciers , and ice fields. Two of the larger

[1] innumerable [i'nju:m(ə)rəb(ə)l] *a.* 无数的，数不清的
[2] creek[kri:k] *n.* 小溪；小湾
[3] pond[pɒnd] *n.* 池塘
[4] glaciated['gleʃi,etid] *a.* 受到冰河作用的；冻结成冰的
[5] Lake Iliamna 伊利亚姆纳湖
[6] Bristol Bay 布里斯托尔湾
[7] tributary['tribjʊt(ə)ri] *n.* 支流；进贡国；附属国
[8] the Yukon Territory 育空地区（加拿大）
[9] the Bering Sea 白令海
[10] Colville 科尔维尔（城市名）
[11] Kuskokwim 卡斯科奎姆（河名）
[12] Noatak 诺阿塔克（河名）
[13] navigable['nævigəb(ə)l] *a.* 可航行的；可驾驶的；适于航行的
[14] accessible[ək'sesib(ə)l] *a.* 易接近的；可进入的；可理解的
[15] sled[slɛd] *n.* 雪橇
[16] navigate['nævigeit] *vt.* 驾驶，操纵；使通过；航行于
[17] salmon['sæmən] *n.* 鲑鱼；大马哈鱼；鲑肉色
[18] tidewater['taidwɔ:tə] *n.* 潮水；低洼海岸；有潮水域
[19] piedmont['pi:dmɒnt] *a.* 山麓的

glaciers in Alaska are Bering and **Malaspina**[20] on the South-central coast , each of which cover almost 2000 square miles ($5180km^2$), an area larger than **Rhode Island**[21] . Most of the glaciers are found in southern Alaska where temperatures are higher than in the north but snowfall is much more abundant. Glaciated areas in the north are smaller and are **confined**[22] to regions of **the Brooks Range**[23].

[20] Malaspina 马拉斯皮纳（冰川名）

[21] Rohle Island 罗德岛

[22] confined[kən'faind] v. 限制（confine 的过去式和过去分词）

[23] the Brooks Range 布鲁克斯岭

Permafrost

Permafrost is an important aspect of Alaska's geography. This is a layer of soil or rock at some depth beneath the surface in which the temperature has been continuously below32°F (0℃) for at least two years.

Permafrost is found where summer heating fails to reach the base of the layer of frozen ground. An active layer that **thaws**[24] during the summer months generally lies **atop**[25] the permanently frozen ground. The active layer, typically less than 3 feet (91cm) in depth, is where the biological activity in the soil takes place. Continuous permafrost covers the northern one third of the state, while the state's southern coastal areas are generally permafrost free. In between lie areas of discontinuous permafrost, where permafrost is often found on north - facing slopes or in poorly drained areas.

[24] thaw[θɔ:] vi. 融解；变暖和

[25] atop[ə'tɒp] prep. 在……的顶上

The most important factor for permafrost is the average annual temperature. In Alaska, the permafrost is colder and deeper with increasing **latitude**[26]. Even though the air temperature can be well below freezing in winter, the permafrost temperature is **moderated**[27] by the **insulating**[28] properties of the seasonal snowpack.

Permafrost is hydrologically important because it acts as an **impenetrable**[29] barrier for water **drainage**[30]. In areas affected by permafrost, construction—must make **accommodations**[31] for these permafrost areas as well as for seasonal freezing and thawing of the ground.

(565words)

[26] latitude['lætitju:d] n. 纬度；界限；活动范围

[27] moderated['mɔdəreitid] v. 缓和，节制（moderate 的过去分词）

[28] insult[in'sʌlt] vt. 侮辱；辱骂；损害

[29] impenetrable[im'penitrəb(ə)l] a. 不能通过的；顽固的；费解的；不接纳的

[30] drainage['dreinidʒ] n. 排水；排水系统；污水；排水面积

[31] accommodation [əkɒmə'deiʃ(ə)n] n. 住处，膳宿；调节；和解；预订铺位

Exercises

1. Build sentences with the following words.

(1) two - thirds, drained, tributaries

(2) permafrost, aspect, Alaska's geography

2. Translate the following sentences into English.

(1) 在阿拉斯加南部发现大多数冰川虽然温度比北部高但是降雪却更频繁。

(2) 影响冻土最重要的因素是年平均温度。

3. Translate the following sentences into Chinese.

(1) These rivers and many other navigable waterways are an important part of Alaska's history, as they provided an essential mode of transportation during the period of early exploration.

(2) Even though the air temperature can be well below freezing in winter, the permafrost temperature is moderated by the insulting properties of the seasonal snowpack.

Learning section

世界著名冰川数量与分布

Antarctica 南极洲 13980000 平方公里
Greenland 格陵兰岛 1802400 平方公里
Arctic Region 北极地区 226090 平方公里
Europe 欧洲 21415 平方公里
Asia 亚洲 109085 平方公里
America 美洲 93022 平方公里

Unit 4. Harbin International Ice and Snow Sculpture Festival

Passage

Harbin is located in Northeast China under the direct influence of the cold winter wind from **Siberia**[1]. The average temperature in summer is 21.2 degrees **Celsius**[2], —16.8 degrees Celsius in winter. Annual low emperatures below —35℃ are not uncommon.

Officially, the festival starts January 5th and lasts one month. However the exhibits often open earlier and stay longer, weather permitting. Ice sculpture decoration technology ranges from the modern (using **lasers**[3]) to traditional (with ice lanterns). While there are ice sculptures throughout the city, there are two main exhibition areas: **Enormous**[4] snow sculptures at Sun Island (a **recreational**[5] area on the opposite side of the Songhua River from the city) and the separate "Ice and Snow World" that operates each night. Ice and Snow World features **illuminated**[6] full size buildings made from blocks of 2 - 3 feet thick **crystal**[7] clear ice directly taken from the Songhua River. There are ice lantern park touring activities held in many parks in the city. Winter activities in the festival include **Yabuli**[8] alpine skiing, winter - swimming in the Songhua River, and the ice - lantern exhibition in **Zhaolin Garden**[9].

The Harbin festival is one of the world's four largest ice and snow festivals, along with Japan's **Sapporo**[10] Snow Festival, Canada's **Quebec**[11] City Winter **Carnival**[12], and Norway's Ski Festival.

(209 words)

[1] Siberia 西伯利亚
[2] celsius['selsiəs] *n.* 摄氏度
[3] laser['leizə(r)] *n.* 激光
[4] enormous[i'nɔ:məs] *a.* 庞大的，巨大的；凶暴的，极恶的
[5] recreational[,rekri'eiʃənl] *a.* 娱乐的，消遣的；休养的
[6] illuminate[i'lu:mineit] *vt.* 阐明，说明；照亮；使灿烂；用灯装饰
[7] crystal['kristl] *a.* 水晶的；透明的，清澈的
[8] Yabuli 亚布力（地名）
[9] Zhaolin Garden 兆麟公园
[10] Sapporo 札幌（日本）
[11] Quebec 魁北克（加拿大）
[12] carnival['kɑ:nivl] *n.* 狂欢节，嘉年华会；饮宴狂欢

Exercises

1. Build sentences with the following words.

(1) Harbin, Northeast China, cold winter wind

(2) festival, starts, lasts

2. Translate the following sentences into English.

（1）然而如果天气允许的话，哈尔滨冰雪节会更早对外开放并且展览持续的时间会更长。

（2）在哈尔滨市的许多公园里会经常举办冰灯游园活动。

3. Translate the following sentences into Chinese.

（1）Winter activities in the festival include Yabuli alpine skiing，winter－swimming in the Songhua River，and the ice－lantern exhibition in Zhaolin Garden.

（2）The Harbin festival is one of the world's four largest ice and snow festivals，along with Japan's Sapporo Snow Festival，Canada's Quebec City Winter Carnival，and Norway's Ski Festival.

Learning section

世界十大最北首都

Reykjavik 冰岛首都雷克雅未克

Helsinki 芬兰首都赫尔辛基

Oslo 挪威首都奥斯陆

Tallinn 爱沙尼亚首都塔林

Stockholm 瑞典首都斯德哥尔摩

Riga 拉脱维亚首都里加

Moscow 俄罗斯首都莫斯科

Copenhagen 丹麦首都哥本哈根

Vilnius 立陶宛首都维尔纽斯

Minsk 白俄罗斯首都明斯克

Unit 5. Groundwater in Permafrost Regions

Passage

In polar latitudes and high mountains, the mean annual temperatures may be sufficiently low for the ground to be at a temperature below 0℃. If this temperature persists for two or more years, the condition is known as permafrost. During the summer, warm temperatures may cause the upper meter or two (3 to 6 feet) of the soil or rock to thaw. This is called the active layer, but underneath it the ground may be frozen to 400 meters (1300 feet).

The **magnitude**[1] of the annual temperature **fluctuation**[2] of the soil is greatest at the surface, and **diminishes**[3] with depth until, at some depth, there is zero annual temperature **amplitude**[4]. The depth at which the maximum annual soil temperature is 0℃ is the permafrost table. In some areas, the permafrost may occur in layers, with zones of unfrozen ground between them. This condition is usually the result of past **climatic**[5] events, and the permafrost distribution is not **congruent**[6] with the present climatic and **thermal**[7] regime. The local depth of permafrost is a function of the **geothermal**[8] **gradient**[9] and the mean annual air temperature.

The insulating cover of glacier ice and large lakes may prevent the formation of permafrost. Permafrost is **likewise**[10] not present beneath the ocean. At high elevations and on north-facing slopes, where the mean temperature is lower, permafrost may be thicker. Lakes and streams also affect the permafrost. Shallow lakes—2 meters (6 feet) deep or less—freeze to bottom in winter and have little effect on the permafrost table, but the permafrost may be warmer; hence, not as thick. Deeper lakes are unfrozen at the bottom and have an insulating effect. Small deep lakes create a saucer-shaped depression in the permafrost table, which, in turn, creates an upward **indentation**[11] in the bottom of the permafrost. Permafrost is absent beneath large deep lakes—

[1] magnitude['mægnitju:d] *n.* 大小；量级；[地震] 震级；重要；光度

[2] fluctuation[ˌflʌktʃʊ'eiʃ(ə)n; -tjʊ-] *n.* 起伏，波动

[3] diminish[di'miniʃ] *vt.* 使减少；使变小

[4] amplitude['æmplitju:d] *n.* 振幅；丰富，充足；广阔

[5] climatic[klai'mætik] *a.* 气候的；气候上的；由气候引起的；受气候影响的

[6] congruent['kɒŋgruənt] *a.* 适合的，一致的；全等的；和谐的

[7] thermal['θɜ:m(ə)l] *a.* 热的，热量的

[8] geothermal [dʒi:ə(ʊ)'θɜ:m(ə)l] *a.* [地物] 地热的；[地物] 地温的

[9] gradient['greidiənt] *n.* [数] [物] 梯度；坡度；倾斜度

[10] likewise['laikwaiz] *ad.* 同样地；也

[11] indentation[inden'teiʃ(ə)n] *n.* 压痕，[物] 刻痕；凹陷；缩排；呈锯齿状

even in the continuous permafrost zone. The effects of large river on distribution of permafrost are similar to those of large deep lakes.

The permafrost table creates **perched**[12] water in the active layer. This results in poorly drained soils and the typical **muskeg**[13] and **marsh**[14] vegetation of tundra regions. Deeper aquifers are recharged only in the absence of permafrost, as the permafrost layer acts a confining layer.

Water below the permafrost layer is confined. The **potentiometric**[15] surface may be in the permafrost layer or even above land surface. Sub－permafrost water may discharge into large rivers and lakes, beneath which unfrozen **conduits**[16] are open. Saline groundwater may exist beneath permafrost, but it is typically not actively circulating. Discharge of groundwater at the surface, especially in winter, may result in the development of sheets of surface ice or large **conical**[17] hills called **pingos**[18]. Pingos have an ice－core and are formed by the upward arch of the ground surface due to hydraulic pressure of groundwater confined by permafrost.

Alluvial[19] river valleys are good sources of groundwater in permafrost regions. The permafrost beneath and along the river may be thin, or absent in places, and alluvial **gravel**[20] **deposits**[21] are good aquifers. Large rivers will have more unfrozen ground; hence, more available water, than smaller tributaries and headwater streams.

In the far northern parts of Alaska, Canada, and **the Soviet Union**[22], permafrost is present nearly everywhere. It is in this continuous permafrost region that maximum permafrost depths are found. To the south, the permafrost layer is discontinuous. It may be up to 180 maters (600 feet) thick locally, but elsewhere there could be unfrozen ground. In the continuous permafrost regions of Alaska, alluvial valleys of large rivers may be the only water available. As the active layer freezes in winter, **base flow**[23] to even the largest rivers may be reduced to zero. Subpermafrost groundwater is typically saline or **brackish**[24]. In the discontinuous permafrost region, the permafrost is thinner, and the areas free of permafrost are much more extensive in alluvial river valleys.

[12] perched[pɜːtʃt] *a.* 栖息的；置于高处的

[13] muskeg[ˈmʌskeg] *n.* （尤指北美北部和北欧的）泥岩沼泽地；青苔沼泽地

[14] marsh[mɑːʃ] *n.* 沼泽；湿地

[15] potentiometric [pəuˌtenʃiəˈmetrik] *a.* 电势测定的，电位计的

[16] conduit[ˈkɒndjuit;ˈkɒndit] *n.* [电] 导管；沟渠；导水管

[17] conical[ˈkɒnik(ə)l] *a.* 圆锥的；圆锥形的

[18] pingo[ˈpiŋgəu] *n.* 小丘

[19] alluvial[əˈluːviəl] *a.* 冲积的

[20] gravel[ˈgræv(ə)l] *n.* 碎石；砂砾

[21] deposit[diˈpɒzit] *n.* 存款；保证金；沉淀物

[22] the Soviet Union 苏联

[23] base flow 基流

[24] brackish[ˈbrækiʃ] *a.* 含盐的；令人不快的；难吃的

Alluvial fans are found in Alaska at the **margins**[25] of many of the mountain ranges. The fans are composed of **glacial- fluvial deposits**[26], which tend to be **coarse**[27] sand and gravel. Groundwater can be obtained from these deposits, either below the permafrost or near rivers or lakes where the permafrost may be thin or absent. In the more **southerly**[28] alluvial fans, the water table may be below the permafrost layer. In this case, the permafrost has little impact on the hydrogeology, other than **restricting**[29] recharge to areas where it is absent.

The distribution of permafrost in **consolidated**[30] rocks is similar to that in unconsolidated deposits. If a rock until has significant hydraulic **conductivity**[31] in both the **horizontal**[32] and **vertical**[33] directions, groundwater hydrology should be similar to that of unconsolidated deposits. In highly **anisotropic**[34] aquifers, even discontinuous permafrost bodies could acts as groundwater dams, preventing horizontal flow and significantly reducing or eliminating vertical recharge. If the **fracture**[35] zone were covered by a **patch**[36] of permafrost, recharge would be difficult, even if the fracture zone extended below the permafrost layer.

(828 words)

[25] margin['mɑ:dʒin] *n.* 边缘；利润，余裕；页边的空白

[26] glacial - fluvial deposit 冰积土

[27] coarse[kɔ:s] *a.* 粗糙的；粗俗的；下等

[28] southerly['sʌðəli] *a.* 来自南方的；向南的

[29] restrict[ri'strikt] *vt.* 限制；约速；限定

[30] consolidated[kən'sɒlideitid] *a.* 巩固的；统一的；整理过的

[31] conductivity[kɒndʌk'tiviti] *n.* 导电性；［物］［生理］传导性

[32] horizontal[hɒri'zɒnt(ə)l] *a.* 水平的；地平线的；同一阶层的

[33] vertical['vɜ:tikl] *a.* 垂直的，直立的；［解剖］头顶的，顶点的

[34] anisotropic [ˌænaisə(ʊ)'trɒpik] *a.* ［物］各向异性的；［物］非均质的

[35] fracture['fræktʃə] *n.* 破裂，断裂；［外科］骨折

[36] patch[pætʃ] *n.* 眼罩；斑点；碎片；小块土地

Exercises

1. Build sentences with the following words.

(1) permafrost, geothermal gradient, annual air temperature

(2) saline groundwater, beneath permafrost, circulating

2. Translate the following sentences into English.

(1) 覆盖在冰川上的冰和大型湖泊也许会防止冻土的形成。

(2) 冻土分布在坚硬的岩石上，类似于疏松的沉积物。

3. Translate the following sentences into Chinese.

(1) Small deep lakes create a saucer - shaped depression in the permafrost table, which, in turn, creates an upward indentation in the bottom of the permafrost.

(2) In this case, the permafrost has little impact on the hydrogeology, other than restricting recharge to areas where it is absent.

Learning section

世界十大滑雪场

Whistler in Canada 加拿大惠斯勒滑雪场
Kitzbuhel in Austria 奥地利基茨比厄尔滑雪场
Zermatt in Switzerland 瑞士策尔马特滑雪场
Vail in USA 美国范尔滑雪场
Banff in Canada 加拿大班夫滑雪场
Chamonix - Mont - Blanc in France 法国霞慕尼滑雪场
Stowe in USA 美国斯托滑雪场
Mont - Tremblant in Canada 加拿大塔伯拉山滑雪场
Cortina in Italy 意大利克缔纳滑雪场
Aspen in USA 美国阿斯本滑雪场

Part 5

Unit 1. Power of Attorney

Passage

China Dongfang Water Conservancy Construction Corp.
ADD: 403 Southern luoshi Road, Wuchang Dist, Wuhan, China
ZIP code: 441300
P. O. Box: Wuhan 777
Fax: (0086 27) 84013135/84014075
Tel: (0086 27) 82381188 (exchange)
E-mail: CDWC@mx. cel. gov. cn.

POWER OF ATTORNEY

Ref. No[1]: CDWC-2002-W1-384
Date: Jun. 18, 2002

KNOW ALL MEN **BY THESE PRESENTS**[2] THAT, I, the undersigned, Liu Dongping, President of China Dongfang Water Conservancy Construction **Corporation**[3] (CDWC in abbreviation), a corporation formed and operating under the laws of the People's Repubic of China, lawfully **authorized**[4] to represent and act **on behalf of**[5] the said Corporation do hereby **appoint**[6] Mr. Chen Dawei, **Vice President**[7] of the said Corporation, whose **signature**[8] appears below, to be its true and lawful **attorney**[9] and authorize the said attorney to **negotiate**[10] Construction Project—Construction of Main Works, Contract No. HH 5.

CDWC admits the legal effect of all his signatures. This Power of **Attorney**[11] will become into **effect**[12] immediately after the signature and be valid until further notice.

[1] ref. No. (reference No) 文件(文件编号)
[2] by these presents 根据本文件
[3] corporation[kɔ:pə'reiʃ(ə)n] n. 公司
[4] authorize['ɔθəraiz] vt. 授权
[5] on behalf of 代表
[6] appoint[ə'pɒint] vt. 任命,指定,授权
[7] Vice President 副总裁,副主席,副董事长
[8] signature ['signətʃə] n. 签字
[9] attorney[ə'tɜ:ni] n. 代理人(被委托人)
[10] negotiate [ni'gəʊʃieit] vt. &vi. 谈判
[11] power of attorney 委托书
[12] become into effect 生效

[13] grantee[grɑːn'tiː] *n.* 被委托者

Grantee[13]	Granter
__________	__________
Chen Dawei	Liu Dongping
Vice President of CDWC	President of CDWC

(153 words)

Learning section

(1) KNOW ALL MEN BY THESE PRESENTS THAT，(法律用语) 根据文件正式地。

(2) I，the undersigned，Liu Dongping，President of… 四个成分属于同位语。

(3) China Dongfang Water Conservancy Construction Corporation 中国东方水利建设公司。

[14] the People's Republic of China 中华人民共和国

(4) a corporation formed and operating under the laws of **the People's Republic of China**[14] 中华人民共和国合法的经营企业。

(5) lawfully authorized to represent and act on behalf of the said Corporation 过去分词短语作定语，修饰注解 (2) 中的四个同位语成分，可以翻译成“公司的法人代表”，the said “上述的，前面所讲的”。

(6) 第一个自然段整个段落为一句话，注解 (2) 中的四个并列的同位语成分作主语，do (hereby) appoint… and authorize 作并列谓语。

(7) Concerned in activities of Huanghu Dam Construction Project - Construction of Main Works，Contract No. HH 5. 过去分词短语作定语修饰 matters。

(8) fish in troubled/muddy waters：混水摸鱼；趁火打劫

get to/reach smooth water：安然摆脱困境；进入顺境

get/wring water from a flint：缘木求鱼

Unit 2. Fax

Passage

World Wide Service One of the largest General Contracting Company in china	中国东方水利建设公司 CHINA DONGFANG WATER CONSERVANCY CONSTRUCTION CORP. HUANGHUN DAM CONSTRUCTION PROJECT，HDR－5
Fax To：Angelica Larios James Instruments Inc. 3727 N. Kedzie Ave. Chicago， IL60618 USA	From：Mr. Li Gaocai CHINA DONGFANG WATER CONSERVANCY CONSTRUCTION CORPORATION（CDWC）
Tel：733. 463. 6565	Date：July 05，1999
Fax：773. 463. 0009 E－mail：info@ndtjames. com	Passage：01 E－mail：Li Gaocai123@163. com Li Gaocai123@sina. com
Subject：**Purchase order**[1]	CC：
□**Urgent**[2] □For **Review**[3] □Please **Comment**[4]	□Please Reply □Please **Recycle**[5]

Dear sir，

We've already remitted you ＄2400. 00 on your specified account for the DATASCAN（R－C－4974）equipment. Please **deliver**[6] the product to us as soon as possible after you receive the **remittance**[7] and fax us informing of that. **Bill of the remittance**[8] is enclosed **hereinafter**[9]. **Best regards**[10]

Mr. Li Gaocai

Project Office：CDWC Office，Huanghun Dam，Wuhan

Tel：(0086) 0598－8838094－7Ext121，122

Fax：(0086) 0598－8838098

[1] purchase order 订货单

[2] urgent[ˈɜːdʒ(ə)nt] *a.* 紧急的

[3] review[riˈvjuː] *n.* &*vt.* &*vi.* 审阅

[4] comment[ˈkɒment] *n.* &*vt.* &*vi.* 点评

[5] recycle [riːˈsaik(ə)l] *n.* &*vt.* &*vi.* 存档

[6] deliver[diˈlivə] *n.* &*vt.* &*vi.* 交货，发货

[7] remittance[riˈmit(ə)ns] *n.* 汇款

[8] bill of the remittance 汇款单

[9] hereinafter[hiərinˈɑːftə] *ad.* 其后，在后，附后

[10] best regards 祝好

Learning section

(1) CHINA DONGFANG WATER CONSERVANCY CONSTRUCTION CORP. “中国东方水利建设公司”。公司名称作“文件头”，常用大写。

(2) HUANGHUN DAM CONSTRUCTION PROJECT，HDR－5“黄昏大坝建设工程5号合同”。详细说明传真的具体地点。

(3) World Wide Service“提供全球服务”。

(4) One of the largest general contracting company in China“中国最大的承包建设公司之一”。

(5) He knows the water the best who has waded through it：要知河深浅，须问过来人

in low water：缺钱，经济拮据

in rough / troubled water：灾难深重

Unit 3. Resume

Passage

1. 英文简历范例

[Name] [Gender][Birth date]
[Street Address],[City, ST/Prov. ZIP Code]
[Phone]
[E-mail]

Objective [求职意向]	*To obtain a Lecturer where I can use my professional, educational and managing skills.*
Skills & Qualifications [技能/资格]	• CET 6(560 p) • IELTS 6.5 • NCRE-Ⅱ (VB) • Driving license B • Violin (level-9) • Java, VF, C^{++} • AutoCAD • GIS • Office 2010 • Photoshop • Speech/Lecturing • Management Skills • Customer Service Skills • Typing (90+ wpm) • SuperWrite (80 wpm)
Education [教育背景]	*China Institute of Water Resources and Hydropower Research, PK, China, Jul. 2011* ***Ph. D.** in **Hydrology and Water Resources*** *Dissertation: "* * * * * * * *"* *Honors: Dissertation passed "with **Distinction**"* *Jilin University Jilin Prov., China, Jun. 2007* ***M. E.** in **Groundwater Science and Engineering*** *Dissertation: "* * * * * * * *"* *Heilongjiang University, Heilongjiang Prov., China, Jun. 2004* ***B. E.** in **Hydropower and Hydraulic Engineering*** *Dissertation: "* * * * * * * *"* *Honors: **G. P. A. 3.8/4.0; Attendance 100%; No. 2 of the department***
Self-description [自我描述]	• *Developed ability to work in a fast-paced atmosphere* • *Maintained excellentteacher-student relations and developed rapport* • *Ability to follow instructions well and make decisions with no supervision* • *Motivated and supervised 15+ employees on daily basis* • *Maintained all record-keeping procedures without error* • *Delegated responsibilities towork to meet company's expectations* • …
Employment History [工作经历]	***AssistantLecturer**, College of Environment and Resources, Jilin University* — *Sep. 2005 - Dec. 2006* …

2. 英文简历常用词汇

(1)个人背景(Tab. 5. 1)。

Tab. 5. 1 个 人 背 景

name	姓名	alias	别名
birth date/place	出生日期、地点	age	年龄
sex/gender	性别	male/female	男/女
native place	籍贯	autonomous region	自治区
province	省	city	市
municipality	直辖市	prefecture	专区
county	县	nationality	民族,国籍
duel citizenship	双重国籍	Address/add.	地址
postal code	邮政编码	Tel/phone No.	电话
height	身高	weight	体重
marital status	婚姻状况	family status	家庭状况
married	已婚	single/unmarried	未婚
divorced	离异	number of children	子女人数
health condition	健康状况	blood type	血型
ID	身份证号码	date of availability	可到职时间

(2)教育程度(Tab. 5. 2)。

Tab. 5. 2 教 育 程 度

(academic) degree	学位	education	学历
ploma	毕业文凭/学位证书	undergraduate	大学生/大学肄业生
(post) graduate	研究生	doctoral student	博士研究生
bachelor	学士	master	硕士
doctor / Ph. D	博士	post doctorate	博士后
senior	大学四年级学生	Junior	大学三年级学生
sophomore	大学二年级学生	freshman	大学一年级学生
guest student	旁听生(英)	intern	实习生
curriculum	课程	social practice	社会实践
major	主修	minor	副修
scholarship	奖学金	Three Goods student	三好学生
student union	学生会	excellent leader	优秀干部
academic year	学年	educational system	学制
semester	学期(美)	term	学期 (英)

(3)个人品质(Tab. 5. 3)。

Tab. 5. 3 **个 人 品 质**

active	主动的,活跃的	adaptable	适应性强的
aggressive	有进取心的	ambitious	有雄心壮志的
considerate	体贴的	cooperative	有合作精神的
dedicated	有奉献精神的	devoted	有献身精神的
dependable	可靠的	diplomatic	老练的,有策略的
disciplined	守纪律的	well - educated	受过良好教育的
energetic	精力充沛的	enthusiastic	充满热情的
expressive	善于表达	faithful	守信的,忠诚的
hospitable	殷勤的	humorous	幽默的
impartial	公正的	independent	有主见的
industrious	勤奋的	initiative	首创精神
intellective	有智力的	intelligent	理解力强的
inventive	有发明才能的	just	正直的
kind - hearted	好心的	knowledgeable	有见识的
learned	精通某门学问的	loyal	忠心耿耿的
modest	谦虚的	motivated	目的明确的
objective	客观的	original	有独创性的
precise	一丝不苟的	persevering	不屈不挠的
punctual	严守时刻的	purposeful	意志坚强的
rational	有理性的	realistic	实事求是的
reasonable	讲道理的	reliable	可信赖的
responsible	负责的	self - conscious	自觉的
selfless	无私的	sincere	真诚的
smart	精明的	spirited	生气勃勃的
steady	踏实的	straightforward	老实的
strong - willed	意志坚强的	sweet - tempered	性情温和的
temperate	稳健的	tireless	孜孜不倦的

Learning section

in smooth water:处于顺境中;日子好过
like water off a duck's back:不起作用;毫无影响
limn on water:瞬刻即逝,徒劳无益

Vocabulary

Vocabularies for Part 1

A

1	accessibility[əkˌsesə'biləti] *n.* 可达性，可及性，可及度	Part 1 - Unit 2
2	accessible[ək'sesəbl] *a.* 允许的，可进入，可达的	Part 1 - Unit 2
3	adjacent[ə'dʒeisənt] *a.* 邻接的，邻近的，邻近（地面）	Part 1 - Unit 5
4	adopt[ə'dɒpt] *vt.* 采用，采取，采纳	Part 1 - Unit 3
5	agricultural[ˌægri'kʌltʃərəl] *a.* 农业的	Part 1 - Unit 7
6	alfalfa [æl'fælfə] *n.* 苜蓿，紫花苜蓿	Part 1 - Unit 5
7	algebra['ældʒibrə] *n.* 代数学，代数	Part 1 - Unit 8
8	alpine['ælpain] *a.* 高山的	Part 1 - Unit 1
9	approximately[ə'prɒksimətli] *ad.* 近似地，大约	Part 1 - Unit 1
10	aquifer['ækwifə(r)]*n.* 地下蓄水层,砂石含水层	Part 1 - Unit 3
11	archive['ɑːkaiv] *n.* 档案，档案馆	Part 1 - Unit 7
12	arid climates 干燥气候	Part 1 - Unit 5
13	asset['æset] *n.* 有价值的人或物，资产，财产	Part 1 - Unit 7
14	assign to 把……分配给	Part 1 - Unit 3
15	atmospheric circulation 大气循环	Part 1 - Unit 2

B

1	balance['bæləns] *n.* 天平	Part 1 - Unit 7
2	base flow 基流	Part 1 - Unit 5
3	beech trees 山毛榉	Part 1 - Unit 5
4	biosphere['baiəʊsfiə(r)]*n.* 生物圈,生命层,生物界	Part 1 - Unit 2
5	biosphere reserves 生物保护圈	Part 1 - Unit 1
6	bone - dry sand dunes 极干的沙丘,砂	Part 1 - Unit 5
7	boreal[bɒriəl] *a.* 北方的	Part 1 - Unit 1
8	branch [brɑːntʃ] *n.* 树枝；分支；部门，分科；支流	Part 1 - Unit 3
9	Bureya *n.* 布列亚河	Part 1 - Unit 1
10	burrowing animals 掘穴动物	Part 1 - Unit 5

Vocabularies for Part 1

C

1	Cape Cod 科德角	Part 1 - Unit 5
2	capillary force 毛管力	Part 1 - Unit 3
3	capillary fringe 毛细管条纹	Part 1 - Unit 3
4	caption['kæpʃn] *n.* 标题，说明文字	Part 1 - Unit 6
5	categorization[ˌkætəgərai'zeiʃn] *n.* 编目方法，分门别类	Part 1 - Unit 3
6	cease[siːs] *vi.* 停止；终了 *vt.* 停止；结束 *n.* 停止	Part 1 - Fig. 1
7	chemical constituent 化学成分，组分，组成成分	Part 1 - Unit 5
8	clay soils 黏性土壤	Part 1 - Unit 5
9	clayey soil 黏质土	Part 1 - Unit 5
10	climatology [ˌklaimə'tɒlədʒi] *n.* 气候学	Part 1 - Unit 8
11	coarse gravel 粗砾，粗砾石，粗卵石	Part 1 - Unit 3
12	coastline['kəustlain] *n.* 海岸线	Part 1 - Unit 2
13	Cold Region 寒区	Part 1 - Unit 8
14	concentration [kəns(ə)n'treiʃ(ə)n] *n.* 集中，浓度，浓缩	Part 1 - Unit 5
15	concrete[kənkriːt] *a.* 具体的，基本的 *n.* 凝结物，愈合，凝香体 *vt.* 凝结，使固结	Part 1 - Unit 5
16	condensation[ˌkɒnden'seiʃn] *n.* 凝结，冷凝	Part 1 - Fig. 1
17	condense [kən'dens] *v.* 浓缩	Part 1 - Fig. 2
18	confluence['kɒnfluəns] *n.* 合流，合流点，集合	Part 1 - Unit 1
19	conservation of mass 质量守恒	Part 1 - Unit 6
20	conservative tracer 防腐剂示踪物	Part 1 - Unit 5
21	continental[ˌkɒnti'nentl] *a.* 大陆的 *n.* 大陆人	Part 1 - Unit 2
22	contour['kɒntuə(r)] *n.* 外形，轮廓，等高线 *a.* 沿等高线修筑的	Part 1 - Fig. 2
23	contribute [kən'triːbjut] *n.* 贡献 *vt.* 有助于 *vi.* 有助于	Part 1 - Unit 5
24	contribute to 导致	Part 1 - Unit 5
25	component[kəm'pəunənt] *n.* 元词，成分，组件 *a.* 组成的	Part 1 - Unit 5
26	CPC：Communist Party of China 中国共产党	Part 1 - Unit 7
27	crane[krein] *n.* 鹤，起重机 *v.* 引颈，伸长（脖子）*vt.* 伸长（脖子等）	Part 1 - Unit 1
28	crust[krʌst] *n.* 外壳，外皮，地壳	Part 1 - Unit 2
29	culpable['kʌlpəbl] *a.* 有罪的	Part 1 - Unit 6
30	curve[kɜːv] *n.* 曲线，弯曲，圆括号 *vt.* 弯曲	Part 1 - Unit 5

D

1	Daurian Steppe Plateau 达乌尔草原高原	Part 1 - Unit 1
2	dean[diːn] *n.* 院长	Part 1 - Unit 7
3	defendant[di'fendənt] *n.* 被告	Part 1 - Unit 6

续表

4	deficit['defisit] *n.* 欠缺，不足	Part 1 - Fig. 2
5	deflect[di'flekt] *vi.* 偏转，偏斜 *vt.* 使偏斜	Part 1 - Unit 5
6	degradation[ˌdegrə'deiʃn] *n.* 降格，堕落，退化	Part 1 - Unit 1
7	deputy['depjuti] *n.* 副手代理人，议员	Part 1 - Unit 7
8	deterioration [diˌtiəriə'reiʃn] *n.* 恶化，退化，变坏	Part 1 - Unit 1
9	deviate['diːvieit] *vi.* 偏离，背离，偏差	Part 1 - Unit 2
10	dike [daik] *n.* 堤，坝	Part 1 - Fig. 2
11	discharge[dis'tʃaːdʒ] *n.* 出院，流量，放射 *vi.* 放电，流出 *vt.* 清偿，排放，放出	Part 1 - Unit 5
12	dissertation[ˌdisə'teiʃn] *n.* 学位论文	Part 1 - Unit 8
13	distribution [ˌdistri'bjuːʃn] *n.* 分发，分配；散布，分布	Part 1 - Fig. 1
14	dominate['dɒmineit] *vt.* 支配，统治	Part 1 - Unit 5
15	drain[drein] *n.* 下水道，排水沟，消耗 *v.* 耗尽，排出沟外	Part 1 - Unit 1
16	drift[drift] *vi.* 漂，漂流；漂泊，流浪 *n.* 漂移，漂流	Part 1 - Fig. 2
17	drilled into 用（机器、工具等）在（某物）上钻孔	Part 1 - Unit 5
18	drivers['draivəz] *n.* 驱动	Part 1 - Unit 2
19	drought[draʊt] *n.* 干旱 *a.* 干旱的	Part 1 - Unit 5
	E	
1	ecology[i'kɒlədʒi] *n.* 生态，生态学	Part 1 - Unit 8
2	elevation[ˌeli'veiʃn] *n.* 海拔，标高，高度	Part 1 - Unit 2
3	endemic[en'demik] *n.* 地方病，风土病 *a.* 风土的，地方的	Part 1 - Unit 1
4	endemic fish 特有鱼种	Part 1 - Unit 7
5	engineering drafting 工程制图	Part 1 - Unit7
6	essentially [i'senʃəli] *ad.* 本质上	Part 1 - Unit 6
7	estate[i'steit] *n.* 土地，财产	Part 1 - Unit 7
8	evaporate [i'væpəreit] *vt.* （使某物）蒸发掉 *vi.* 消失，不复存在	Part 1 - Fig. 2
9	evaporation[iˌvæpə'reiʃn] *n.* 蒸发（作用）	Part 1 - Fig. 1
10	evapotranspiration[iˌvæpəʊˌtrænspi'reiʃən] *n.* 土壤水分蒸发蒸腾损失总量	Part 1 - Unit 5
11	excess[ik'ses] *n.* 过量，过剩	Part 1 - Fig. 2
12	exponential decay 衰变指数	Part 1 - Unit 5
13	extensive[iks'tensiv] *a.* 广泛的，广阔的	Part 1 - Unit 1
14	extraction[ik'strækʃn] *vt.* 提取	Part 1 - Unit 1
	F	
1	Far East 远东	Part 1 - Unit 7
2	finer - grained 细粒的	Part 1 - Unit 3

续表

3	flat topography 地势平坦	Part 1 - Unit 5
4	fluctuate['flʌktʃʊeit] *vi.* 波动，起伏 *vt.* 拨动	Part 1 - Unit 2
5	flux[flʌks] *n.* 通量，焊剂，焊剂 *vi.* 流动，*vt.* 出水	Part 1 - Unit 2
6	foothill['futhil] *n.* 山麓小丘	Part 1 - Unit 1
7	frame[freim] *n.* 帧，框架，架	Part 1 - Unit 2
8	function ['fʌŋkʃn] *n.* 函项，函数，作用	Part 1 - Unit 5
	G	
1	gage[geidʒ] *n.* 抵押，规，表	Part 1 - Unit 5
2	general hydrogeology 水文地质学基础	Part 1 - Unit 8
3	geography [dʒi'ɒgrəfi] *n.* 地理（学）	Part 1 - Unit 8
4	geoscience[ˌdʒiːəʊ'saiəns] *n.* 地球科学	Part 1 - Unit 9
5	geotechnical[dʒiːəu'teknikəl] *a.* 土力学的	Part 1 - Unit 7
6	geothermal[ˌdʒi(ː)əu'θəməl] *a.* 地热的，地温的	Part 1 - Unit 7
7	glacier['glæsiə(r)] *n.* 冰川，冰河	Part 1 - Unit 2
8	glacier ice 冰川冰	Part 1 - Unit 2
9	glassware['glɑːswɛə] *n.* 玻璃器具类	Part 1 - Unit 7
10	grandeur ['grændʒə] *n.* 富丽堂皇	Part 1 - Unit 1
11	ground ice 地下冰	Part 1 - Unit 7
12	groundwater dynamics 地下水动力学	Part 1 - Unit 8
13	groundwater volume 地下水容积	Part 1 - Unit 2
14	groundwater ['graʊndwɔːtə (r)] *n.* 地下水	Part 1 - Fig. 2
	H	
1	headwater['hedwɔːtə(r)] *n.* 上游源头	Part 1 - Unit 1
2	Henty 亨提山	Part 1 - Unit 2
3	Great Hinggan 大兴安岭	Part 1 - Unit 2
4	Stanovoy 外兴安岭	Part 1 - Unit 1
5	heritage['heritidʒ] *n.* 遗产，继承物	Part 1 - Unit 1
6	horizontally [ˌhɒri'zɒntəli] *ad.* 水平	Part 1 - Unit 5
7	humidity [hjuː'midəti] *n.* 湿度，潮湿，湿气	Part 1 - Unit 5
8	Hydraulic & Hydropower Engineering 水利水电工程	Part 1 - Unit 7
9	hydraulics[hai'drɔːliks] *n.* 水力学	Part 1 - Unit 7
10	hydrochemical[haidrəʊ'kemikl] *a.* 水化学的	Part 1 - Unit 7
11	hydrograph ['haidrəgræf] *n.* 水文曲线，水位图	Part 1 - Unit 5

续表

12	hydrography[hai'drɒgrəfi] *n.* 水文地理学	Part 1 - Unit 8
13	hydrologic[ˌhaidrə'lɔdʒik] *a.* 水文的，水文学的	Part 1 - Fig. 1
14	hydrologic balance 水文均衡	Part 1 - Unit 6
15	hydrologic cycle 水循环	Part 1 - Fig. 2
16	hydrological statistics 水文统计	Part 1 - Unit 8
17	hydrologist[hai'drɒlədʒist] *n.* 水文学家	Part 1 - Fig. 2
18	hydrology[hai'drɒlədʒi] *n.* 水文学. hydro -（前缀）水文的	Part 1 - Unit 7
19	hydrology textbook 水文学教材	Part 1 - Unit 5
20	hydrometry[hai'drɒmitri] *n.* 水文测验	Part 1 - Unit 8
	I	
1	impermeable[im'pɜːmiːəbəl] *a.* 不能渗透的，不透水，不渗透的	Part 1 - Unit 5
2	in reference to 关于	Part 1 - Unit 3
3	incentive[in'sentiv] *n.* 刺激，诱因，动机 *a.* 激励的	Part 1 - Unit 5
4	incompressible [inkəm'presib(ə)l] *a.* 非压缩的，不可压缩的，不能压缩的	Part 1 - Unit 6
5	Indiana[ˌindi'ænə] *n.* 印地安那州	Part 1 - Unit 5
6	infiltrate ['infiltreit] *vt.* & *vi.*（使）渗透，（使）渗入	Part 1 - Unit 5
7	inject[in'dʒekt] *vt.* 注射，喷射	Part 1 - Fig. 2
8	injection well 注入井	Part 1 - Unit 5
9	inspirational[ˌinspə'reʃənl] *a.* 鼓舞人心的；带有灵感的，给予灵感的	Part 1 - Fig. 1
10	institute['institjuːt] *n.* 协会，学会，学院，研究院	Part 1 - Unit 9
11	Institute of Groundwater in Cold Region 寒区地下水研究所	Part 1 - Unit 7
12	intensity[in'tensəti] *n.* 强度	Part 1 - Fig. 2
13	intercept[ˌintə'sept] *n.* 遮断，截距，截取 *vt.* 侦听，截取，截听	Part 1 - Unit 5
14	interflow[ˌintə(ː)'fləu] *n.* 混流，层间流，土内水流 *vi.* 混流，合流	Part 1 - Unit 5
15	international environmental agreements 国际环境协定	Part 1 - Unit 1
16	iodide['aiədaid] *n.* 碘化物	Part 1 - Fig. 2
17	silver iodide 碘化银	Part 1 - Fig. 2
18	irregularly[i'regjələli] *a.* 不整齐的，不规则的，不规则的 *n.* 不合规格之物	Part 1 - Unit 2
19	irrigation [ˌiri'geiʃn] *n.* 灌溉	Part 1 - Unit 7
	K	
1	Khanka *n.* 兴凯湖	Part 1 - Unit 1
	L	
1	leopard['lepəd] *n.* 豹	Part 1 - Unit 1
2	lowland['ləulænd] *n.* 低地	Part 1 - Unit 1

续表

3	lumped[lʌmpt] *a.* 集总的 *vt.* 集总	Part 1 - Unit 2
4	lysimeter [lai'simitə] *n.* 渗水计，测渗计，渗漏测定计	Part 1 - Unit 5
	M	
1	Maine *n.* 缅因州	Part 1 - Unit 5
2	markedly['mɑːkidli] *ad.* 显著地	Part 1 - Unit 5
3	marsh[mɑːʃ] *n.* 沼泽，湿地，草沼	Part 1 - Unit 2
4	Massachusetts *n.* 马萨诸塞州	Part 1 - Unit 5
5	medium. *n.* 媒质，介质	Part 1 - Unit 3
6	meteorological factors *n.* 气象因素	Part 1 - Unit 5
7	meteorologist[ˌmiːtiə'rɒlədʒist] *n.* 气象学家	Part 1 - Fig. 2
8	meteorology[ˌmiːtiə'rɒlədʒi] *n.* 气象学	Part 1 - Unit 8
9	millimeter ['miliˌmiːtə] *n.* 毫米	Part 1 - Unit 3
10	mineral['minərəl] *a.* 矿物的 *n.* 矿物，矿石	Part 1 - Unit 1
11	mineral surfaces 矿物表面	Part 1 - Unit 4
12	mining['mainiŋ] *n.* 采矿（业）	Part 1 - Unit 9
13	modification[ˌmɒdifi'keiʃn] *n.* 改善，改变	Part 1 - Fig. 2
14	moist[mɔist] *a.* 潮湿的，微湿的，湿润的	Part 1 - Fig. 2
15	Mongolian [mɔŋ'guliən] *n.* 属于蒙古人种的人，蒙古语，蒙古症患者	Part 1 - Unit 1
16	monsoon[ˌmɒn'suːn] *n.* 季风，季节风	Part 1 - Unit 2
17	municipal[mjʊ'nisip(ə)l] *a.* 市政的，市的；地方自治的	Part 1 - Unit 5
18	municipality [mju'nisə'pæləti] *n.* 市区	Part 1 - Unit 5
	N	
1	national nature conservation programs 国家自然保护项目	Part 1 - Unit 1
2	net flux 流网	Part 1 - Unit 2
3	nominated['nɔmineitid] *a.* 被提名的，被任命的	Part 1 - Unit 1
4	nonfiction [nən'fikʃən] *n.* 非小说文学	Part 1 - Unit 6
5	nonpumping well 非抽水井	Part 1 - Unit 5
6	North Korea 朝鲜	Part 1 - Unit 2
7	Northeast Asia 东北亚	Part 1 - Unit 1
	O	
1	oak trees 橡树，栎	Part 1 - Unit 5
2	occur [ə'kɜː] *vi.* 发生；出现	Part 1 - Unit 5
3	onset['ɒnset] *n.* 波至，攻击，发动	Part 1 - Unit 5
4	operating ['ɔpəreitiŋ] *a.* 运行的 *n.* 操作，运转	Part 1 - Unit 6
5	originate [ə'ridʒineit] *vt.* 发源，创造，发自 *n.* 起始，起源	Part 1 - Unit 5
6	overland flow 表面径流，漫流	Part 1 - Unit 5

P

1	particle[ˈpɑːtikl] *n.* 微粒，颗粒，[物] 粒子；极	Part 1 - Fig. 2
2	permeable[ˈpɜːmiːəbəl] *a.* 可渗透的，具渗透性的	Part 1 - Unit 4
3	phreatic zone 地下水层；地下水区；潜水带；饱和层	Part 1 - Unit 3
4	piece[piːs] *n.* 块，片，段；部分，部件；文章，音乐作品 *vt.* 修补；连接，接上	Part 1 - Unit 6
5	plain[plein] *n.* 平原	Part 1 - Unit 1
6	plateau[ˈplætəu] *n.* 高原，平稳，稳定状态	Part 1 - Unit 1
7	plowing[plauiŋ] *n.* 耕地	Part 1 - Fig. 2
8	plus[plʌs] *a.* 正的；*n.* 正号，加 *prep.* 加 *vt.* 加	Part 1 - Unit 5
9	polytechnic[ˌpɒliˈteknik] *n.* 理工学院	Part 1 - Unit 9
10	pore water pressure 孔隙水压力	Part 1 - Unit 3
11	porous and permeable soil 多孔的（具导管的，有孔的）渗透性土壤	Part 1 - Unit 5
12	portion[ˈpɔːʃn] *n.* 部分，分与财产，一部分	Part 1 - Unit 5
13	postgraduate[ˌpəustˈgrædʒuət] *n.* 研究生 *a.* 研究生的	Part 1 - Unit 7
14	practicum[ˈpræktikəm] *n.* 实习科目，实习课	Part 1 - Unit 8
15	precipitation[priˌsipiˈteiʃn] *n.* 降水	Part 1 - Fig. 1
16	predict[priˈdikt] *vt.* 预测	Part 1 - Fig. 2
17	preheat[ˌpriːˈhiːt] *vt.* 预热	Part 1 - Unit 7
18	pressure tank 压力槽，高压锅，耐压试验水筒，坞室	Part 1 - Fig. 1
19	puddle[ˈpʌdl] *n.* 水坑，熔池 *vi.* 和泥浆	Part 1 - Unit 5
20	Pump & Pumping Station 水泵与泵站	Part 1 - Unit 7
21	pump[pʌmp] *n.* 泵，心脏，抽运	Part 1 - Unit 5
22	pumping wells 抽油井	Part 1 - Unit 6
23	purification[ˌpjʊərifiˈkeiʃn] *n.* 净化	Part 1 - Fig. 2

R

1	Ramsar wetlands 拉姆萨尔湿地（《拉姆萨尔公约》挑选出的具有国际重要意义的湿地）	Part 1 - Unit 1
2	recede [riˈsiːd] *vt.* 后退，贬值 *vi.* 后退	Part 1 - Unit 5
3	recession[riˈseʃ(ə)n] *n.* 后退，撤回，不景气	Part 1 - Unit 5
4	recharge[viːˈtʃaːdʒ] *vt.* 再充电，再控告 *n.* 再充电，补给	Part 1 - Unit 5
5	reduce to 把……降低到	Part 1 - Unit 5
6	reservoir[ˈrezəvwɑː(r)] *n.* 水库；储藏，汇集；大量的储备；储液	Part 1 - Fig. 2
7	residence time 停留时间	Part 1 - Unit 2
8	result in 导致，引起	Part 1 - Unit 3
9	retard[riˈtaːd] *vt.* 停滞 *vi.* 减慢；受到阻滞 *n.* 减速；阻滞；延迟	Part 1 - Fig. 2

续表

10	root[ruːt] *n.* 根，根部	Part 1 - Fig. 2
11	rotting tree 腐败，腐朽的树木	Part 1 - Unit 5
12	roughly['rʌfli] *ad.* 概略地	Part 1 - Unit 2
13	route[ruːt] *n.* 路程，道路；途径，渠道槽	Part 1 - Fig. 2
14	run off[rʌn ɔf] 径流	Part 1 - Fig. 2
	S	
1	saline['seilain] *n.* 盐泉，盐水，盐湖 *a.* 含盐的，盐的，苦涩的	Part 1 - Unit 2
2	saturated['sætʃəreitid] *a.* 饱和的；浸透的；(颜色)未被白色弄淡的	Part 1 - Unit 3
3	schematic [skiː'mætik] *a.* 纲要的，示意的，概要的	Part 1 - Unit 5
4	screened['skriːnd] *a.* 过筛的	Part 1 - Unit 5
5	seepage['siːpidʒ] *n.* 渗流	Part 1 - Unit 7
6	separately['seprətli] *ad.* 独立地	Part 1 - Unit 2
7	septic system 腐败性系统	Part 1 - Unit 5
8	setting['setiŋ] *n.* 背景，装置，栽植	Part 1 - Unit 5
9	sewage water 污水	Part 1 - Unit 5
10	shaded['ʃeidid] *a.* 荫蔽的	Part 1 - Unit 5
11	shower['ʃauə(r)] *n.* 阵雨，淋浴，淋洒器 *vt.* 浇灌	Part 1 - Unit5
12	shrink[ʃriŋk] *vt.* & *vi.* 收缩，皱缩	Part 1 - Unit 2
13	significantly [sig'nifikəntli] *ad.* 意义重大地	Part 1 - Unit 2
14	silt[silt] *n.* 淤泥，泥沙沉积，粉砂	Part 1 - Unit 3
15	simulation[ˌsimju'leiʃn] *n.* 模拟，模仿	Part 1 - Unit 7
16	siphon['saifən] *n.* 虹吸管，虹吸 *vt.* 吮吸	Part 1 - Unit 5
17	sleet[sliːt] *n.* 雨夹雪或雹 *vi.* 下雨夹雪；下冻雨	Part 1 - Fig. 2
18	sloping['sləupiŋ] *a.* 倾斜的，有坡度的	Part 1 - Fig. 2
19	soggy rain forest 湿润的雨林，常雨乔木群落	Part 1 - Unit 5
20	soggy soil 湿润土壤	Part 1 - Unit 5
21	soil water 土壤水	Part 1 - Unit 3
22	solar energy 太阳能	Part 1 - Unit 2
23	solar radiation 太阳辐射	Part 1 - Unit 2
24	spill over 从……溢出	Part 1 - Unit 5
25	stabilize ['steibilaiz] *vt.* 稳定 *vi.* 安定	Part 1 - Unit 3
26	statistics[stə'tistiks] *n.* 统计学	Part 1 - Unit 8
27	steppe[step] *n.* 特指西伯利亚一带没有树木的大草原	Part 1 - Unit 1
28	stochastic[stə'kæstik] *a.* 随机的	Part 1 - Unit 8

续表

29	storage['stɔːridʒ] *n.* 贮存，贮藏	Part 1 - Fig. 2
30	storage tank 储水罐	Part 1 - Fig. 1
31	stork[stɔːk] *n.* [鸟] 鹳	Part 1 - Unit 1
32	streamflow['striːmfləu] *n.* 流速及流水量	Part 1 - Fig. 1
33	streams[st'riːmz] *n.* 流（stream 的名词复数）	Part 1 - Fig. 2
34	sublimation [ˌsʌbli'meiʃn] *n.* 升华，升华作用	Part 1 - Fig. 2
35	subscript['sʌbskript] *n.* 下标，添标，索引	Part 1 - Unit 5
36	subsoil['sʌbsɔil] *n.* 下土层，底土	Part 1 - Fig. 2
37	sufficient transpiration 充分蒸发	Part 1 - Unit 5
38	surficial[sə'fiʃəl] *a.* 地表的，地面的	Part 1 - Unit 5
39	sustain[sə'stein] *vt.* 维持，使……生存下去	Part 1 - Fig. 2
40	synonymous [si'nɒniməs] *a.* 同义的；同义词的	Part 1 - Unit 3
	T	
1	Texas['teksəs] *n.* 德克萨斯州	Part 1 - Unit 6
2	taiga['teigə] *n.* 针叶树林地带	Part 1 - Unit 1
3	tank[tæŋk] *n.* 振荡回路，筒，柜	Part 1 - Unit 5
4	tap[tæp] *n.* 丝椎，水龙头，分接头 *vi.* 攻丝，轻叩	Part 1 - Unit 5
5	technically['teknikli] *ad.* 技术上；学术上；专业上；严格说来	Part 1 - Unit 3
6	terminology[ˌtəːmi'nɒlədʒi] *n.* 专门名词；术语，术语学	Part 1 - Unit 4
7	terrestrial[ti'restriəl] *n.* 地球上的人	Part 1 - Unit 1
8	territory['teritəri] *n.* 领土，版图，领域，范围	Part 1 - Unit 1
9	theoretical[ˌθiə'retikl] *a.* 理论的	Part 1 - Unit 5
10	timber['timbə] *n.* 木材，木料	Part 1 - Unit 1
11	transgress[træns'gres] *vt.* 违犯，违反	Part 1 - Unit 6
12	transmit[træns'mit] *vt.* 传送，透射，传播	Part 1 - Unit 5
13	transpiration[ˌtrænspi'reʃən] *n.* 蒸发，散发；[植] 蒸腾作用；[航] 流逸	Part 1 - Fig. 1
14	transpiration [ˌtrænspi'reiʃn] *n.* 蒸发（物），散发，蒸腾作用，流逸	Part 1 - Fig. 2
15	triggered['trigəd] *a.* 触发的，起动的	Part 1 - Unit 1
16	tundra['tʌndrə] *n.* 苔原，冻土地带	Part 1 - Unit 1
	U	
1	ultimately['ʌltimətli] *ad.* 最后	Part 1 - Unit 2
2	unconfined[ˌʌnkən'faind] *a.* 无限制的，非承压的	Part 1 - Unit 5
3	undergo[ʌndə'gəʊ] *vt.* 经历	Part 1 - Unit 5
4	UNESCO—United Nations Educational, Scientific and Cultural Organization 联合国教育科学文化组织	Part 1 - Unit 1

续表

5	uniqueness [ju'niknis] *n.* 独特性	Part 1 - Unit 1
6	unsaturated['ʌn'sætʃəreitid] *a.* 没有饱和的，不饱和的	Part 1 - Unit 3
7	urban['ɜːbən] *a.* 城市的	Part 1 - Fig. 1
8	Ussuri[u'suːri] *n.* 乌苏里江	Part 1 - Unit 1
	V	
1	vadose zone 渗流区	Part 1 - Unit 3
2	vapor['veipə] *n.* 水汽，水蒸气 *v.* 自夸，(使) 蒸发	Part 1 - Fig. 2
3	vegetation [ˌvedʒə'teiʃn] *n.* 植物 (总称)；草木	Part 1 - Fig. 2
4	vice[vais] *a.* 副的	Part 1 - Unit 7
5	vice versa 反之亦然	Part 1 - Unit 2
6	virtually['vɜːtʃuəli] *ad.* 事实上	Part 1 - Unit 2
	W	
1	wade [weid] *vt.* & *vi.* (从水、泥等) 蹚，走过；跋涉 *n.* 跋涉	Part 1 - Fig. 1
2	wastage['weistidʒ] *n.* 废物，损失，消耗	Part 1 - Unit 5
3	Water Conservancy 水利	Part 1 - Unit 7
4	water table 地下水位	Part 1 - Unit 3
5	water content 含水量，含水率	Part 1 - Unit 5
6	watershed['wɔːtəʃed] *n.* 流域	Part 1 - Fig. 2
7	well discharge 井流量	Part 1 - Unit 5
	Y	
1	yields[jiːld] *n.* 产量，产额	Part 1 - Unit 6
	Z	
1	Zeya *n.* 泽亚河	Part 1 - Unit 1

Vocabularies for Part 2

A

1	abutment [ə'bʌtmənt] *n.* 邻接，桥墩，桥基；扶垛；对接；接界	Part 2 - Unit 2
2	afterbay['ɑːftəbei] *n.* 尾水池	Part 2 - Fig. 1
3	alternate ['ɔːltəːnət] *a.* 轮流的；交替的；间隔的；代替的	Part 2 - Unit 2
4	alternative [ɔːl'təːnətiv] *a.* 非正统的，不寻常的；两者择一的	Part 2 - Unit 3
5	arch [ɑːtʃ] *n.* 弓形；拱门；拱形物；足弓，齿弓	Part 2 - Unit 2
6	attributes [ə'tribjuːts] *n.* 属性，特性，特质；属性把……归于；把……品质归于某人	Part 2 - Unit 3

B

1	blade[bleid] *n.* 刀刃，叶片	Part 2 - Fig. 3
2	budget['bʌdʒit] *n.* 预算；预算额，经费 *vt.* & *vi.* 编制预算，安排开支等	Part 2 - Unit 7
3	buttress ['bʌtrəs] *n.* 扶壁；支撑物 *vt.* 支持，鼓励；用扶壁支撑，加固	Part 2 - Unit 2

C

1	cross section 横截面	Part 2 - Fig. 1
2	canyon ['kænjən] *n.* 峡谷	Part 2 - Unit 2
3	cavitation[ˌkævi'teiʃən] *n.* 空洞形成，气穴现象；空化；涡凹；汽蚀	Part 2 - Unit 5
4	chute [ʃuːt] *n.* 斜槽，滑道；降落伞 *vt.* 用斜槽或斜道运送 *vi.* 顺斜道而下，	Part 2 - Unit 5
5	coincidence[kəu'insidəns] *n.* 符合；一致；同时发生	Part 2 - Unit 1
6	cold region 寒区	Part 2 - Unit 7
7	compact['kɒmpækt] *v.* 压紧，（使）坚实	Part 2 - Unit 4
8	composite ['kɒmpəzit] *a.* 混合成的，［建］综合式的；［数］可分解的 *n.* 合成物	Part 2 - Unit 4
9	conservative[kən'səːvətiv] *n.* 保守的人；（英国）保守党党员，保守党支持者	Part 2 - Unit 5
10	coolant ['kuːlənt] *n.* 冷冻剂，冷却液，散热剂	Part 2 - Unit 4
11	costly['kɔstli] *a.* 昂贵的，代价高的，花钱多的；引起困难的；造成损失的	Part 2 - Unit 1
12	counteract [ˌkaʊntər'ækt] *vt.* 抵消；阻碍；中和	Part 2 - Unit 3
13	crest[krest] *n.* 山顶；羽毛饰；鸡冠；（动物的）颈脊	Part 2 - Unit 5
14	cut - off trench 齿墙	Part 2 - Fig. 1

D

1	definition [ˌdefi'niʃn] *n.* 定义；规定，明确；［物］清晰度；解释	Part 2 - Unit 6
2	detention basin 蓄洪区，拦洪区	Part 2 - Unit 1
3	dike[daik] *n.* 堤；排水沟；障碍物	Part 2 - Unit 2
4	discharge [dis'tʃɑːdʒ] *n.* 流出，泄流	Part 2 - Unit 1

续表

5	dissimilar [di'similə (r)] *a*. 不同的，不相似的	Part 2 - Unit 4
6	diversion[dai'vɜːʃn] *n*. 转移；分散注意力；消遣	Part 2 - Unit 2
7	downstream[ˌdaun'striːm] *ad*. 在下游，顺流地	Part 2 - Fig. 1
	E	
1	eliminate [i'limineit] *vt*. 消（排，清）除	Part 2 - Unit 1
2	embankment [im'bæŋkmənt] *n*. 路堤；筑堤	Part 2 - Unit 2
3	entrepreneur[ˌɔntrəprə'nəː] *n*. 企业家，创业，创业者	Part 2 - Unit 7
	F	
1	flexibility[ˌfleksi'biliti] *n*. 柔韧性；机动性，灵活性，易曲性；适应性，弹性	Part 2 - Unit 1
2	flood crest 洪峰，同 flood peak	Part 2 - Unit 1
3	floodgate['flʌdgeit] *n*. （江河或湖泊的）防洪闸（门），（泄）水闸门	Part 2 - Unit 2
4	flood - mitigation reservoir 减洪水库，滞洪水库	Part 2 - Unit 1
5	frost resistance 抗冻性	Part 2 - Unit 7
	G	
1	generator['dʒenəreitə] *n*. 发电机，发生器	Part 2 - Fig. 2
2	generator shaft 发电机轴	Part 2 - Fig. 3
3	geosynthetic[ˌdʒiə'sin'θetik]] *n*. 土工合成材料	Part 2 - Unit 7
4	gravity dam 重力坝	Part 2 - Fig. 1
5	granular['grænjələ(r)] *a*. 颗粒状的	Part 2 - Unit 4
	H	
1	homogeneous[ˌhɒmə'dʒiːniəs] *a*. 同性质同类的；由相同（或同类型）事物组成的	Part 2 - Unit 4
2	hydroelectric['haidrəui'lektrik] *a*. 水力发电的	Part 2 - Fig. 2
	I	
1	impervious[im'pɜːviəs] *a*. 不可渗透的；透不过的；无动于衷的；不受影响的	Part 2 - Unit 3
2	impound[im'paʊnd] *vt*. 将……关起来；扣押，监禁；搁置，保留；蓄水	Part 2 - Unit 2
3	infamous['infəməs] *a*. 声名狼藉的；无耻的，伤风败俗的	Part 2 - Unit 5
4	inflow['infləu] *n*. 进水量，流入，入流	Part 2 - Unit 1
5	Inner Mongolia 内蒙古自治区	Part 2 - Unit 8
6	installation[ˌinstə'leiʃən] *n*. 安装，设置；就职；装置，设备	Part 2 - Unit 1
7	intake['inteik] *n*. 吸入，纳入，（液体等）进入口，进水口	Part 2 - Fig. 2
8	intricate ['intrikət] *a*. 错综复杂的；难理解的；曲折；盘错	Part 2 - Unit 2
	K	
1	kilowatt ['kiləwɒt] *n*. [电]千瓦	Part 2 - Unit 6
2	kinetic[ki'netik] *a*. 运动的，活跃的，能动的，有力的；[物] 动力（学）的，运动的	Part 2 - Unit 6

	L	
1	levee['levi] *n.* 堤；早朝	Part 2 - Unit 2
2	liquefaction [ˌlikwi'fækʃən] *n.* 液化	Part 2 - Unit 4
	M	
1	masonry['meisənri] *n.* 石工工程，砖瓦工工程；砖石建筑	Part 2 - Unit 4
2	mechanics[mi'kæniks] *n.* 力学	Part 2 - Unit 7
3	megawatt ['megəwɒt] *n.* 兆瓦，百万瓦特（电能计量单位）	Part 2 - Unit 6
4	membrane ['membrein] *n.*（动物或植物体内的）薄膜；隔膜；膜状物	Part 2 - Unit 4
5	migration[mai'greiʃn] *n.* 迁移，移居	Part 2 - Unit 4
6	mitigation[ˌmiti'geiʃən] *n.* 缓解，减轻，平静	Part 2 - Unit 1
7	motivator ['məutiveitə (r)] *n.* 激起行为的人（或事物），促进因素，激发因素	Part 2 - Unit 3
	N	
1	negligence['neglidʒəns] *n.* 疏忽，粗心大意	Part 2 - Unit 1
	O	
1	ogee ['əudʒiː] *n.* 弯曲，S形；表反曲线	Part 2 - Unit 5
2	orifice['ɔrifis] *n.* 孔口，管口	Part 2 - Unit 1
3	outflow['autfləu] *n.* 流出量；放水，出流	Part 2 - Unit 1
4	outlet['autlet] *n.* 出口，出路，出水口	Part 2 - Fig. 2
5	outweigh[aut'wei] *vt.* 重于，比……（重要）	Part 2 - Unit 1
6	overtop['əuvə'tɒp] *vt.* 高出，高耸……之上	Part 2 - Unit 5
	P	
1	penstock['penstək] *n.* 水道，水渠，压力水管，水阀门	Part 2 - Fig. 2
2	penstock ['penstɒk] *n.*〈美〉水道；水渠；压力水管；水阀门	Part 2 - Unit 6
3	permafrost['pəːməfrɔst] *n.*（如极地的）永久冻土；多年冻土	Part 2 - Unit 4
4	permanent['pəːmənənt] *a.* 永久（性）的，固定的；永恒的；长久的	Part 2 - Unit 1
5	pile [pail] *v.* 堆起；堆叠；放置；装入	Part 2 - Unit 3
6	pinwheel ['pinwiːl] *n.* 轮转焰火，纸风车；针轮；靶中心	Part 2 - Unit 6
7	power transmission cable 电力传输电缆	Part 2 - Fig. 2
8	powerhouse['pauəˌhaus] *n.* 发电厂房	Part 2 - Fig. 2
9	predetermine[ˌpriːdi'təːmin]*vt.* & *vi.* 预先裁定；注定	Part 2 - Unit 5
10	prohibitive[prə'hibətiv] *a.* 禁止的；禁止性的；抑制的；（指价格等）过高的	Part 2 - Unit 4
11	provision [prə'viʒn] *n.* 规定，预备，准备 *vt.* & *vi.* 为……提供所需物品（指食物）	Part 2 - Unit 2
	R	
1	rampart['ræmpɑːt] *n.*（城堡等周围宽阔的）防御土墙；防御，保护	Part 2 - Unit 2
2	recover[ri'kʌvə] *vt.* 恢复；重新获得，找回	Part 2 - Unit 1

续表

3	recreation [ˌriːkriˈeiʃn] *n.* 消遣（方式）；娱乐（方式）；重建，重现	Part 2 - Unit 6
4	Reinforced Concrete Structure 钢筋混凝土结构	Part 2 - Unit 7
5	release [riˈliːs] *vt.* 释放；放开；发布；发行 *n.* 释放，排放，解除；释放令	Part 2 - Unit 6
6	represent [ˌrepriˈzent] *vt.* 表现，象征；代表，代理；扮演；作为示范 *vi.* 代表；提出	Part 2 - Unit 3
7	reservoir[ˈrezəvwɑː] *n.* 水库，蓄水体	Part 2 - Fig. 1
8	retarding basin 滞洪区	Part 2 - Unit 1
9	rotor[ˈrəutə] *n.* 转子	Part 2 - Fig. 3

S

1	safe capacity of the channel downstream 下游河道安全泄水量	Part 2 - Unit 1
2	scouring[ˈskauəriŋ] *n.* 擦（洗）净，冲刷，洗涤，（用力）刷；擦净；擦亮	Part 2 - Unit 5
3	shaft[ʃɑːft] *n.* 柄，轴；矛，箭；光线 *vt.* 给……装上杆柄	Part 2 - Unit 5
4	silt[silt] *n.* 淤泥	Part 2 - Fig. 2
5	siphon[ˈsaifn] *n.* 虹吸管 *vt.* 用虹吸管吸（或输送）（液体） *vi.* 通过虹吸管	Part 2 - Unit 5
6	sloping [ˈsləupiŋ] *a.* 倾斜的，有坡度的 *v.* 悄悄地走	Part 2 - Unit 2
7	sluice gate 泄水闸门	Part 2 - Fig. 2
8	sluice [sluːs] *n.* 水闸；有闸人工水道；漂洗处 *vt.* 冲洗；漂净；给……安装水闸	Part 2 - Unit 2
9	sluiceway[ˈsluːsˌwei] *n.* 泄洪道，分洪道	Part 2 - Unit 1
10	spall [spɔːl] *n.* （尤指岩石的）碎片，裂片 *vt.* & *vi.* 弄碎，击碎（矿石）	Part 2 - Unit 2
11	spillway[ˈspilwei] *n.* 溢洪道，泄洪道	Part 2 - Unit 1
12	stator[ˈsteitə] *n.* 定子，固定片	Part 2 - Fig. 3
13	storage capacity 蓄水库容，库容	Part 2 - Unit 1
14	submerged[səbˈməːdʒd] *a.* 在水中的，淹没的	Part 2 - Unit 1
15	susceptible [səˈseptəbl] *a.* 易受影响的；易受感染的；善感的；可以接受或允许的	Part 2 - Unit 4

T

1	temporarily[ˈtempθrərili] *ad.* 暂时地	Part 2 - Unit 1
2	tender[ˈtendə] *n.* 投标 *a.* 脆弱的，幼弱的；温柔的，亲切的；疼痛的，敏感的	Part 2 - Unit 7
3	thawing [ˈθɔːiŋ] *n.* 熔化 *v.* （气候）解冻；缓和；溶化	Part 2 - Unit 2
4	Tibet[tiˈbet] *n.* 西藏	Part 2 - Unit 8
5	timber [ˈtimbə（r）] *n.* 木材，木料；用材林，林场；素质 *vt.* 用木料支撑；备以木材	Part 2 - Unit 4
6	transformer[trænsˈfɔːmə] *n.* 变压器	Part 2 - Fig. 2
7	trench[trentʃ] *n.* 深沟，地沟；战壕	Part 2 - Fig. 1
8	trough[trɒf] *n.* 水槽，食槽；低谷期；[航] 深海漕；[气] 低气压槽	Part 2 - Unit 5
9	turbine blade 涡轮叶片	Part 2 - Fig. 3
10	turbine [ˈtəːbain] *n.* 涡轮机；汽轮机；透平机	Part 2 - Unit 6

U

1	ungated[ʌn'geitid] *a.* 无门的；无闸门的	Part 2 - Unit 1
2	upstream[ˌʌp'striːm] *ad.* &*a.* 向上游地（的），逆流地（的）	Part 2 - Fig. 1

V

1	valve[vælv] *n.* 阀，活门；（心脏的）瓣膜；真空管	Part 2 - Unit 1

W

1	watertight ['wɔːtətait] *a.* 不漏水的；水密的；防渗的；无懈可击的	Part 2 - Unit 4
2	wicket['wikit] *n.* 三柱门	Part 2 - Fig. 3
3	wicket gate 导叶	Part 2 - Fig. 3

Vocabularies for Part 3

A

1	accompany [ə'kʌmpæni] v. 伴随……同时发生；陪伴；陪同	Part 3 - Unit 1
2	adaptation[ædəp'teiʃ(ə)n] n. 适应，顺应；同化	Part 3 - Unit 2
3	aggregate['ægrigət] vt. 使聚集，使积聚	Part 3 - Unit 5
4	agronomic[ˌægrə'namik] a. 农事的；农艺学的	Part 3 - Unit 5
5	algal['ælgəl] a. 海藻的	Part 3 - Unit 3
6	alignment[ə'lainm(ə)nt] n. 队列，成直线；校准；结盟	Part 3 - Unit 6
7	apparently[ə'pærəntli] ad. 显然地；表面上；显而易见	Part 3 - Unit 1
8	aqueduct['ækwidʌkt] n.[水利] 渡槽；导水管；沟渠	Part 3 - Unit 5
9	arable['ærəb(ə)l] a. 适用于耕种的 n. 耕地	Part 3 - Unit 1
10	arid['ærid] a. 干旱的，干燥的；贫瘠的，荒芜的，不毛的	Part 3 - Unit 5
11	asphalt['æsfælt] n. 沥青，柏油；（铺路等用的）沥青混合料	Part 3 - Unit 5
12	avocado[ˌævə'kado] n. 鳄梨，鳄梨树；暗黄绿色	Part 3 - Unit 3

B

1	basin irrigation 淹灌	Part 3 - Unit 2
2	bonus['bəunəs] n. 奖金，额外津贴；红利；额外令人高兴的事情	Part 3 - Unit 3
3	border['bɔːdə] n. 畦；边；镶边；边界	Part 3 - Unit 2
4	border irrigation 畦灌	Part 3 - Unit 2
5	brackish['brækiʃ] a. 有盐味的，可厌的	Part 3 - Unit 5
6	branch[brɑːn(t)ʃ]n. 树枝，分枝；分部；支流 vi. &vt. 分支；出现分歧	Part 3 - Unit 3
7	bubblers['bʌb(ə)lə(r)] n. 喷水式饮水口	Part 3 - Unit 3

C

1	carbon['kaːbən] n. 碳	Part 3 - Fig. 1
2	carbon dioxide 二氧化碳	Part 3 - Fig. 1
3	category['kætig(ə)ri] n. 种类，类别；派别	Part 3 - Unit 1
4	coarse[kɔːs]a. 粗糙的；粗鄙的	Part 3 - Unit 2
5	Coastal and Offshore Engineering 海岸与近海工程	Part 3 - Unit 7
6	complex['kɔmpleks] a. 由许多部分组成的，复合的；复杂的，难懂的	Part 3 - Unit 7
7	component [kəm'pəunənt] n. 成分；组分；零件	Part 3 - Unit 3
8	comprehensive[kɒmpri'hensiv] n. 综合学校；a. 综合的	Part 3 - Unit 4
9	conjunctive [kən'dʒʌŋ (k) tiv] a. 连接的	Part 3 - Unit 5

续表

10	consecutive[kən'sekjutiv]a. 连续的，连贯的	Part 3 - Unit 1
11	conservancy[kən'səːvənsi] n.（自然物源的）保护，管理，水土保持	Part 3 - Unit 7
12	control box 首部枢纽	Part 3 - Unit 3
13	conventional irrigation 传统灌溉	Part 3 - Fig. 2
14	corrugation[ˌkɔːru'geiʃən] n. 沟；车辙	Part 3 - Unit 2
15	coupling['kʌpliŋ]n.［电］耦合；结合，联结	Part 3 - Unit 5
16	culverts['kʌlvəts] n.［交］涵洞；暗沟（culvert 的复数）	Part 3 - Unit 5

D

1	devices[di'vaisis]n.［机］［计］设备；［机］装置；［电子］器件（device 的复数）	Part 3 - Unit 5
2	dioxide[dai'ɔksaid] n. 二氧化物	Part 3 - Fig. 1
3	discipline['disiplin] vt. 训练，训导；处罚，惩罚	Part 3 - Unit 7
4	discounted['diskauntid] a. 已折扣的 v. 打折扣；不重视	Part 3 - Unit 4
5	disposal[di'spəuz(ə)l] a. 处理（或置放）废品的	Part 3 - Unit 6
6	ditch[ditʃ] n. 沟渠；壕沟	Part 3 - Unit 2
7	doctoral degree program 博士学位授予点，博士学位学科点	Part 3 - Unit 7
8	downstream ['dɑun'striːm] a. 下游的；顺流的 ad. 下游地；顺流而下	Part 3 - Unit 6
9	dragline['dræglain] n. 牵引绳索；［机］拉铲挖土机；绳斗电铲	Part 3 - Unit 6
10	drainage['dreinidʒ]n. 排水，放水；排水系统，下水道；废水，污水，污物	Part 3 - Unit 8
11	drawdown['drɔːdaun] n.（抽水后）水位降低，水位降低量	Part 3 - Fig. 2
12	dredge[dredʒ] n. 挖泥船，疏浚机；拖捞网	Part 3 - Unit 6
13	drip irrigation 滴灌	Part 3 - Fig. 2
14	drought[draut] n. 旱灾；干旱（时期），旱季	Part 3 - Unit 1
15	drought - resistant 抗旱的	Part 3 - Unit 5
16	dynamic[dai'næmik] a. 有活力的，强有力的；不断变化的；动力的，动态的	Part 3 - Fig. 2

E

1	edge[edʒ] n. 边缘；锋利，尖锐	Part 3 - Unit 3
2	elevation[ˌeli'veiʃən] n. 海拔	Part 3 - Fig. 2
3	eliminate[i'limineit] vt. 排除，消除；淘汰，除掉	Part 3 - Unit 3
4	emitter[i'mitə] n. 发射器；这里指滴灌灌水器滴头	Part 3 - Unit 3
5	encrustation [ɛn'krʌs'teʃən] n. 硬壳；结壳；用覆盖物	Part 3 - Unit 3
6	enhance[in'hæns] vt. 提高；增加；加强	Part 3 - Unit 1
7	erosion[i'rəuʒ(ə)n] n. 侵蚀，腐蚀，磨损	Part 3 - Unit 3
8	estuarine['estjuərin] a. 河口的，江口的	Part 3 - Unit 7
9	excavated['ɛkskəˌvetid] v. 发掘；挖掘（excavate 的过去式，过去分词）	Part 3 - Unit 6

续表

10	extension[ik'stenʃ(ə)n] *n.* 延长；延期；扩大；伸展；电话分机	Part 3 - Unit 4
11	extreme[ik'stri:m] *a.* 极端的，过激的；极限的，非常的	Part 3 - Unit 3
	F	
1	facility[fə'siliti] *n.* 设备；设施；能力	Part 3 - Unit 1
2	feature['fi:tʃə] *n.* 特征，特点；容貌，面貌	Part 3 - Unit 6
3	fertilizer['fɜ:tilaizə] *n.* 肥料，化肥	Part 3 - Unit 3
4	filter['filtə] *vt.* 过滤；透过；渗透	Part 3 - Unit 3
5	furrow irrigation 沟灌	Part 3 - Unit 2
	G	
1	guarantee[gær(ə)n'ti:] *vt.* 保证；担保	Part 3 - Unit 1
	H	
1	ha[hɑ:] *n.* 面积公顷（hectare）缩写，在我国常使用 hm^2	Part 3 - Unit 3
2	harbor['ha:bə] *n.* 海港	Part 3 - Unit 7
3	herbicide['hɜ:bisaid] *n.* 除草剂	Part 3 - Unit 3
4	hydraulic drops 水力降；[水利] 跌水	Part 3 - Unit 5
5	hydrologic [hɑid'rəulədʒik] *n.* 水文学	Part 3 - Unit 6
	I	
1	injury['in(d)ʒ(ə)ri] *n.* 伤害；损害；伤害的行为	Part 3 - Unit 2
2	instability[instə'biliti] *n.* 不稳定（性）；基础薄弱；不安定	Part 3 - Unit 6
3	interfere[intə'fiə] *vi.* 干预，干涉；调停，排解；妨碍，打扰	Part 3 - Unit 6
	L	
1	lateral['læt(ə)r(ə)l] *a.* 侧面的，横向的，这里 lateral line 指毛管	Part 3 - Unit 3
2	laterals['læt(ə)r(ə)ls] *n.* 侧根；侧面部分（lateral 的复数）	Part 3 - Unit 6
3	lieu[lju:; lu:] *n.* 代替；场所，处所	Part 3 - Unit 6
4	limnology[lim'nɔlədʒi] *n.* 湖泊学	Part 3 - Unit 7
5	liter ['litər] *n.* 公升（容量单位）	Part 3 - Unit 3
6	loans[lons] *n.* [金融] 贷款（loan 的复数形式）；借贷	Part 3 - Unit 4
	M	
1	maize[meiz] *n.* 玉米；黄色，玉米色	Part 3 - Unit 2
2	municipal[mjʊ'nisip(ə)l] *a.* 市的，市政的；地方自治的；都市的	Part 3 - Unit 6
3	nozzle['nɒz(ə)l] *n.* 管嘴，喷嘴	
	O	
1	objectionable[əb'dʒekʃ(ə)nəb(ə)l] *a.* 令人不快的，讨厌的	Part 3 - Unit 6
2	orchard['ɔ:tʃəd] *n.* 果园；果园里的全部果树	Part 3 - Unit 3
3	oxygen['ɔksidʒən] *n.* 氧，氧气	Part 3 - Fig. 1

P

1	paddy['pædi] *n.* 稻田（复数 paddies）	Part 3 - Unit 4
2	parallel['pærəlel] *n.* 平行线（面）；相似物	Part 3 - Unit 2
3	pasture['pɑːstʃə] *n.* 牧草地，牧场	Part 3 - Unit 2
4	pedology[pi'dələdʒi] *n.* 土壤学	Part 3 - Unit 8
5	perennial[pə'reniəl] *a.* 终年的，长久的；多年生的	Part 3 - Unit 1
6	permeability[pɜːmiə'biliti] *n.* 渗透性；磁导率；可渗透性	Part 3 - Unit 2
7	photosynthesis[ˌfəutəu'sinθisis] *n.* 光合作用，光能合成	Part3 - Fig. 1
8	plowing[plauiŋ] *n.*［农学］翻耕，耕作 *v.* 耕地；犁	Part 3 - Unit 4
9	plug[plʌg] *vi.* 填塞，堵	Part 3 - Unit 3
10	porous['pɔːrəs] *a.* 多孔渗水的；能渗	Part 3 - Unit 3
11	precipitation[pri,sipi'teiʃ(ə)n] *n.* 降雨量	Part 3 - Unit 1
12	Pump and Pumping Station 水泵与泵站	Part 3 - Unit 8

Q

1	quota['kwəutə] *n.* 配额；定额；限额	Part 3 - Unit 5

R

1	rationally['ræʃənli] *ad.* 讲道理地，理性地	Part 3 - Unit 5
2	regulate['regjʊleit] *vt.* 调节，调整；控制，管理	Part 3 - Unit 3
3	rehabilitate[riːhə'biliteit] *vt.* 使康复；使恢复原状 *vi.* 复兴；复权	Part 3 - Unit 5
4	ridge[ridʒ] *n.* 田埂；背脊，峰；隆起线；山脊	Part 3 - Unit 2
5	row[rəʊ] *n.* 行；排；路，街；吵闹	Part 3 - Unit 2
6	runoff['rʌnɒf] *n.*［水文］径流；决赛；流走的东西 *a.* 决胜的	Part 3 - Unit 6

S

1	schemes[skiːmz] *n.* 计划（scheme 的名词复数）；体系；阴谋	Part 3 - Unit 1
2	sediment['sedim(ə)nt] *n.* 沉积物，沉渣	Part 3 - Unit 3
3	seepage['siːpidʒ] *n.*［流］渗流；渗漏；渗液	Part 3 - Unit 4
4	semi - humid ['semihjuːmid] *n.* 半潮湿，半湿性	Part 3 - Unit 1
5	siphon['saif(ə)n] *n.* 虹吸管	Part 3 - Unit 2
6	spatial['speiʃ(ə)l] *a.* 空间的；存在于空间的；受空间条件限制的	Part 3 - Unit 5
7	State Key Laboratory 国家重点实验室	Part 3 - Unit 7
8	steeper [s'tiːpə] *n.* 浸润器；浸泡用的桶子 *a.* 陡峭的，险峻的	Part 3 - Unit 6
9	strip[strip] *n.* 长条；带状地带	Part 3 - Unit 2
10	supplemental[ˌsʌpli'mentəl]*a.* 补充的，追加的	Part 3 - Unit 1

T

1	temporal['temp(ə)r(ə)l] *a.* 时间的；世俗的；暂存的	Part 3 - Unit 5

续表

2	terraced['terəst] *a*. 阶地的；*v*. 使成阶地	Part 3 - Unit 4
3	tillage['tilidʒ] *n*. 耕作，耕种	Part 3 - Unit 5
4	topography[tə'pɒgrəfi] *n*. 地形；地貌；地形学	Part 3 - Unit 2
5	trapezoidal[ˌtræpi'zɔidəl] *a*. ［数］梯形的；不规则四边形的	Part 3 - Unit 6
6	turbulence['tə:bjuləns] *n*. 气体或水的涡流；波动	Part 3 - Unit 7
	U	
1	undergraduate[ˌʌndə'grædjuit] *n*.（未获学士学位的）大学生，大学肄业生	Part 3 - Unit 7
2	unlined[ˌʌn'laind] *a*. ［服装］无衬里的	Part 3 - Unit 6
3	unsightly[ʌn'saitli] *a*. 不美观，难看的，不好看的	Part 3 - Unit 6
4	urban['ɜ:b(ə)n] *a*. 都市的	Part 3 - Unit 5
	V	
1	vineyard['vinjɑ:d] *n*. 葡萄园	Part 3 - Unit 3
	W	
1	water spreading 漫灌	Part 3 - Unit 2

Vocabularies for Part 4

A

1	absorption[əb'zɔːpʃ(ə)n] *n.* 吸收；全神贯注，专心致志	Part 4 - Unit 1
2	accessible[ək'sesib(ə)l] *a.* 易接近的；可进入的；可理解的	Part 4 - Unit 3
3	accommodation[əkɒmə'deiʃ(ə)n] *n.* 住处，膳宿；调节；和解	Part 4 - Unit 3
4	Adak 埃达克（美国）	Part 4 - Unit 1
5	adverse['ædvɜːs] *a.* 不利的；相反的；敌对的	Part 4 - Unit 2
6	aided['eidid] *v.* 帮助（aid 的过去分词）	Part 4 - Unit 1
7	Alaskan 阿拉斯加州人的	Part 4 - Unit 1
8	albedo[æl'biːdəʊ] *n.*（行星等的）反射率；星体反照率	Part 4 - Unit 1
9	Aleutians 阿留申群岛	Part 4 - Unit 1
10	alluvial[ə'luːviəl] *a.* 冲积的	Part 4 - Unit 5
11	alternately[ɔːl'tɜːnətli] *ad.* 交替地；轮流地；隔一个地	Part 4 - Unit 2
12	amplitude['æmplitjuːd] *n.* 振幅；丰富，充足；广阔	Part 4 - Unit 5
13	anchor['æŋkə] *vt.* 抛锚；使固定；主持节目	Part 4 - Unit 1
14	anisotropic[ˌænaisə(ʊ)'trɒpik] *a.* ［物］各向异性的；［物］非均质的	Part 4 - Unit 5
15	atop[ə'tɒp] *prep.* 在……的顶上	Part 4 - Unit 3

B

1	Barrow 巴罗（美国）	Part 4 - Unit 1
2	base flow 基流	Part 4 - Unit 5
3	brackish['brækiʃ] *a.* 含盐的；令人不快的；难吃的	Part 4 - Unit 5
4	Bristol Bay 布里斯托尔湾	Part 4 - Unit 3

C

1	carnival['kɑːnivl] *n.* 狂欢节，嘉年华会；饮宴狂欢	Part 4 - Unit 4
2	celsius['selsiəs] *n.* 摄氏度	Part 4 - Unit 4
3	Chita 赤塔	Part 4 - Unit 2
4	climatic[klai'mætik] *a.* 气候的；气候上的；由气候引起的；受气候影响的	Part 4 - Unit 5
5	climatology[klaimə'tɒlədʒi] *n.* 气候学；风土学	Part 4 - Unit 1
6	coarse[kɔːs] *a.* 粗糙的；粗俗的；下等的	Part 4 - Unit 5
7	Cold Bay 科尔德弯（美国）	Part 4 - Unit 1
8	Colville 科尔维尔（城市名）	Part 4 - Unit 3
9	compaction[kəm'pækʃən] *n.* 压紧；精简；密封；凝结	Part 4 - Unit 1

续表

10	conductivity[kɒndʌk'tiviti] *n.* 导电性；[物][生理]传导性	Part 4 - Unit 5
11	conduit['kɒndjʊit; 'kɒndit] *n.* [电]导管；沟渠；导水管	Part 4 - Unit 5
12	confined[kən'faind] *v.* 限制（confine 的过去式和过去分词）	Part 4 - Unit 3
13	congruent['kɒŋgrʊənt] *a.* 适合的，一致的；全等的；合谐的	Part 4 - Unit 5
14	conical['kɒnik(ə)l] *a.* 圆锥的；圆锥形的	Part 4 - Unit 5
15	consecutive[kən'sekjʊtiv] *a.* 连贯的；连续不断的	Part 4 - Unit 1
16	consolidated[kən'sɒlideitid] *a.* 巩固的；统一的；整理过的	Part 4 - Unit 5
17	creek[kriːk] *n.* 小溪；小湾	Part 4 - Unit 3
18	crystal['kristl] *a.* 水晶的；透明的，清澈的	Part 4 - Unit 4

D

1	density['densəti] *n.* 密度	Part 4 - Unit 1
2	deposit[di'pɒzit] *n.* 存款；保证金；沉淀物	Part 4 - Unit 5
3	diminish[di'miniʃ] *vt.* 使减少；使变小	Part 4 - Unit 5
4	drainage['dreinidʒ] *n.* 排水；排水系统；污水；排水面积	Part 4 - Unit 3

E

1	enormous[i'nɔːməs] *a.* 庞大的，巨大的；凶暴的，极恶的	Part 4 - Unit 4

F

1	fluctuation[ˌflʌktʃʊ'eiʃ(ə)n] *n.* 起伏，波动	Part 4 - Unit 5
2	forum['fɔːrəm] *n.* 论坛，讨论会；法庭；公开讨论的广场	Part 4 - Unit 2
3	fracture['fræktʃə] *n.* 破裂，断裂；[外科]骨折	Part 4 - Unit 5

G

1	generality[dʒenə'ræliti] *n.* 概论；普遍性；大部分	Part 4 - Unit 2
2	geocryology [dʒiːəukri'ɔlədʒi] *n.* 冻土地貌学	Part 4 - Unit 2
3	geothermal[dʒiːə(ʊ)'θɜːm(ə)l] *a.* [地物]地热的；[地物]地温的	Part 4 - Unit 5
4	glacial - fluvial deposit 冰积土	Part 4 - Unit 5
5	glaciated['gleʃiˌetid] *a.* 受到冰河作用的；冻结成冰的	Part 4 - Unit 3
6	gradient['greidiənt] *n.* [数][物]梯度；坡度；倾斜度	Part 4 - Unit 5
7	gravel['græv(ə)l] *n.* 碎石；砂砾	Part 4 - Unit 5

H

1	hemisphere['hemisfiə] *n.* 半球	Part 4 - Unit 2
2	horizontal[hɒri'zɒnt(ə)l] *a.* 水平的；地平线的；同一阶层的	Part 4 - Unit 5

I

1	illuminate[i'luːmineit] *vt.* 阐明，说明；照亮；使灿烂；用灯装饰	Part 4 - Unit 4
2	impenetrable[im'penitrəb(ə)l] *a.* 不能通过的；顽固的；费解的；不接纳的	Part 4 - Unit 3

续表

3	indentation[indenˈteiʃ(ə)n] *n*. 压痕，[物] 刻痕；凹陷；缩排；呈锯齿状	Part 4 - Unit 5
4	innumerable[iˈnjuːm(ə)rəb(ə)l] *a*. 无数的，数不清的	Part 4 - Unit 3
5	insult[inˈsʌlt] *vt*. 侮辱；辱骂；损害	Part 4 - Unit 3
6	interior[inˈtiəriə] *n*. 内部；本质	Part 4 - Unit 1

J

1	Juneau 朱诺（美国）	Part 4 - Unit 1

K

1	Ketchikan 凯奇坎（美国）	Part 4 - Unit 1
2	Kuskokwim 卡斯科奎姆（河）	Part 4 - Unit 3

L

1	Lake Iliamna 伊利亚姆纳湖	Part 4 - Unit 3
2	laser[ˈleizə(r)] *n*. 激光	Part 4 - Unit 4
3	latitude[ˈlætitjuːd] *n*. 纬度；界限；活动范围	Part 4 - Unit 3
4	likewise[ˈlaikwaiz] *ad*. 同样地；也	Part 4 - Unit 5

M

1	magnitude[ˈmægnitjuːd] *n*. 大小；量级；[地震] 震级；重要；光度	Part 4 - Unit 5
2	maintenance[ˈmeintənəns] *n*. 维护，维修；保持；生活费用	Part 4 - Unit 2
3	Malaspina 马拉斯皮纳（冰川名）	Part 4 - Unit 3
4	margin[ˈmɑːdʒin]*n*. 边缘;利润,余裕;页边的空白	Part 4 - Unit 5
5	maritime[ˈmæritaim] *a*. 海的；海事的；沿海的；海员的	Part 4 - Unit 1
6	marsh[mɑːʃ] *n*. 沼泽；湿地	Part 4 - Unit 5
7	metamorphosis[ˌmetəˈmɔːfəsis] *n*. 变形；变质	Part 4 - Unit 1
8	Mirny 米尔内（萨哈共和国）	Part 4 - Unit 2
9	moderated[ˈmɔdəreitid] *v*. 缓和，节制（moderate 的过去分词）	Part 4 - Unit 3
10	muskeg[ˈmʌskeg] *n*.（尤指北美北部和北欧的）泥岩沼泽地；青苔沼泽地	Part 4 - Unit 5

N

1	navigable[ˈnævigəb(ə)l] *a*. 可航行的；可驾驶的；适于航行的	Part 4 - Unit 3
2	navigate[ˈnævigeit] *vt*. 驾驶，操纵；使通过；航行于	Part 4 - Unit 3
3	Nenana 尼纳纳（美国）	Part 4 - Unit 1
4	Noatak 诺阿塔克（河）	Part 4 - Unit 3

O

1	onset[ˈɒnset] *n*. 开始，着手；发作；攻击，进攻	Part 4 - Unit 1

P

1	patch[pætʃ] *n*. 眼罩；斑点；碎片；小块土地	Part 4 - Unit 5

续表

2	perched[pɜtʃt] *a.* 栖息的；置于高处的	Part 4 - Unit 5
3	periglacial[ˌperiˈgleiʃ(ə)l] *n.* 冰缘，冰边	Part 4 - Unit 2
4	piedmont[ˈpiːdmɒnt] *a.* 山麓的	Part 4 - Unit 3
5	pingo[ˈpiŋgəʊ] *n.* 小丘	Part 4 - Unit 5
6	pond[pɒnd] *n.* 池塘	Part 4 - Unit 3
7	potentiometric[pəuˌtenʃiəˈmetrik] *a.* 电势测定的，电位计的	Part 4 - Unit 5

Q

1	Quebec 魁北克（加拿大）	Part 4 - Unit 4

R

1	radiation[reidieiʃ(ə)n] *n.* 辐射；发光；放射物	Part 4 - Unit 1
2	recall[riˈkɔːl] *n.* 召回；回忆；撤销	Part 4 - Unit 1
3	recreational[ˌrekriˈeiʃənl] *a.* 娱乐的，消遣的；休养的	Part 4 - Unit 4
4	restrict[riˈstrikt] *vt.* 限制；约束；限定	Part 4 - Unit 5
5	Rohle Island 罗德岛	Part 4 - Unit 3

S

1	salmon[ˈsæmən] *n.* 鲑鱼；大马哈鱼；鲑肉色	Part 4 - Unit 3
2	Sapporo 札幌（日本）	Part 4 - Unit 4
3	Siberia 西伯利亚	Part 4 - Unit 4
4	signify[ˈsignifai] *vt.* 表示；意味；预示	Part 4 - Unit 1
5	sled[slɛd] *n.* 雪橇	Part 4 - Unit 3
6	southerly[ˈsʌðəli] *a.* 来自南方的；向南的	Part 4 - Unit 5
7	symposium[simˈpəuziəm] *n.* 讨论会，座谈会；专题论文集；酒宴，宴会	Part 4 - Unit 2

T

1	thaw[θɔː] *vi.* 融解；变暖和	Part 4 - Unit 3
2	the Bering Sea 白令海	Part 4 - Unit 3
3	the Brooks Range 布鲁克斯岭	Part 4 - Unit 3
4	the Copper River Basin 科珀河流域	Part 4 - Unit 1
5	Valdez 瓦尔迪兹（美国）	Part 4 - Unit 1
7	the Southeast Panhandle 阿拉斯加东南的狭长地带	Part 4 - Unit 1
8	the Soviet Union 苏联	Part 4 - Unit 5
9	the Tanana River 塔纳纳河	Part 4 - Unit 1
10	the Yukon Territory 育空地区（加拿大）	Part 4 - Unit 3
11	thermal[ˈθɜːm(ə)l] *a.* 热的，热量的	Part 4 - Unit 5
12	tidewater[ˈtaidwɔːtə] *n.* 潮水；低洼海岸；有潮水域	Part 4 - Unit 3

续表

13	transition[træn'ziʃ(ə)n] *n.* 过渡；转变；[分子生物] 转换；变调	Part 4 - Unit 1
14	tributary['tribjut(ə)ri] *n.* 支流；进贡国；附属国	Part 4 - Unit 3
15	tripod['traipɒd] *n.* [摄] 三脚架；三脚桌	Part 4 - Unit 1
	V	
1	vertical['vɜːtikl] *a.* 垂直的，直立的；[解剖] 头顶的，顶点的	Part 4 - Unit 5
	Y	
1	Yabuli 亚布力（地名）	Part 4 - Unit 4
2	Yakutsk 雅库茨克	Part 4 - Unit 2
	Z	
1	Zhaolin Garden 兆麟公园	Part 4 - Unit 4

Vocabularies for Part 5

A

1	appoint[ə'pɒint] *vt.* 任命，指定，授权	Part 5 - Unit 1
2	attorney[ə'tɜːni] *n.* 代理人（被委托人）	Part 5 - Unit 1
3	authorize['ɔθəraiz] *vt.* 授权	Part 5 - Unit 1

B

1	become into effect 生效	Part 5 - Unit 1
2	best regards 祝好	Part 5 - Unit 2
3	bill of the remittance 汇款单	Part 5 - Unit 2
4	by these presents 根据本文件	Part 5 - Unit 1

C

1	comment['kɒment] *n.* &*vt.* *vi.* 点评	Part 5 - Unit 2
2	corporation[kɔːpə'reiʃ(ə) n] *n.* 公司	Part 5 - Unit 1
3	deliver[di'livə] *n.* &*vt.* *vi.* 交货，发货	Part 5 - Unit 2

G

1	grantee[grɑːn'tiː] *n.* 被委托者	Part 5 - Unit 1

H

1	hereinafter[hiərin'ɑːftə] *ad.* 其后，在后，附后	Part 5 - Unit 2
2	negotiate[ni'gəuʃieit] *vt.* &*vi.* 谈判	Part 5 - Unit 1

O

1	on behalf of 代表	Part 5 - Unit 1

P

1	power of attorney 委托书	Part 5 - Unit 1
2	purchase order 订货单	Part 5 - Unit 2

R

1	recycle [ri'saik (ə) l] *n.* &*vt.* &*vi.* 存档	Part 5 - Unit 2
2	ref. No. (reference No) 文件（文件编号）	Part 5 - Unit 1
3	remittance[ri'mit(ə)ns] *n.* 汇款	Part 5 - Unit 2
4	review[ri'vjuː] *n.* &*vt.* &*vi.* 审阅	Part 5 - Unit 2

S

1	signature['signətʃə] *n.* 签字	Part 5 - Unit 1

T

1	the People's Republic of China 中华人民共和国	Part 5 - Unit 1

U

1	urgent['ɜːdʒ(ə)nt] *a*. 紧急的	Part 5 - Unit 2

V

1	Vice President 副总裁，副主席，副董事长	Part 5 - Unit 1

参 考 文 献

[1] 迟道才，周振民．水利专业英语 [M]．北京：中国水利水电出版社，2006.

[2] 张道真．大学英语语法 [M]．济南：山东科学技术出版社，2008.

[3] Anastasiya Markina. The Amur - Heilong River Basin [EB/OL]. http://amur - heilong. net/, 2010.

[4] Charles R. Fitts. Groundwater Science [M]. ACADEMIC PRESS, 2001.

[5] Martha Shulski and Gerd Wender. The Climate of Alaska [M]. Fairbanks: University of Alaska, 2007.

[6] Neil Davis. PERMAFROST [M]. Fairbanks: University of Alaska, 2001.

[7] 陈天照．水利水电工程实用英语 [M]．北京：中国电力出版社，2006.

[8] C. W. FETTER, JR. Applied Hydrogeology [M]. Columbus: Charles E. Merrill Publishing Co, 1980.

[9] 杨雅璘，付强．农业水土工程英语 [M]．哈尔滨：哈尔滨工程大学出版社，2003.

[10] Rural Water Resources Department, Ministry of Water Resources, P. R. China, China Irrigation and Drainage Development Center. China Water Saving Irrigation [M]. BeiJing: China Waterpower Press.

[11] Wkipedia. Dam. [online]. Available from: http://en. wikipedia. org/wiki/Dam, 2013. 5.

[12] Wkipedia. Hydroelectric _ power. [online]. Available from: http://en. wikipedia. org/wiki/Hydroelectric _ power, 2013. 5.